Springer Series in
Electronics and Photonics 30

Edited by D. H. Auston

Springer Series in Electronics and Photonics

Editors: D. H. Auston   W. Engl   T. Sugano

Managing Editor: H. K. V. Lotsch

This series was originally published under the title
Springer Series in Electrophysics
and has been renamed starting with Volume 22.

Volumes 1–20 are listed on the back inside cover

H.M. Gibbs   G. Khitrova
N. Peyghambarian (Eds.)

Nonlinear Photonics

With 133 Figures

Springer-Verlag
Berlin Heidelberg New York London
Paris Tokyo Hong Kong Barcelona

Dr. Hyatt M. Gibbs
Dr. Galina Khitrova
Dr. Nasser Peyghambarian

Optical Sciences Center, University of Arizona,
Tucson, AZ 85721, USA

Series Editors:

Dr. David H. Auston

Columbia University, Dept. of Electrical Engineering, New York, NY 10027, USA

Professor Dr. Walter Engl

Institut für Theoretische Elektrotechnik, Rhein.-Westf. Technische Hochschule,
Templergraben 55, D-5100 Aachen, Fed. Rep. of Germany

Professor Takuo Sugano

Department of Electronic Engineering, The Faculty of Engineering,
The University of Tokyo, 7-3-1, Hongo, Bunkyo-ku, Tokyo, 113, Japan

Managing Editor: Dr. Helmut K. V. Lotsch

Springer-Verlag, Tiergartenstrasse 17,
D-6900 Heidelberg, Fed. Rep. of Germany

ISBN-13: 978-3-642-75440-1   e-ISBN-13: 978-3-642-75438-8
DOI: 10.1007/978-3-642-75438-8

Library of Congress Cataloging-in-Publication Data. Nonlinear photonics / H. M. Gibbs, G. Khitrova, N. Peyghambarian (eds.). p. cm.–(Springer series in electronics and photonics ; v. 30) Includes bibliographical references.   1. Photonics. 2. Nonlinear optics. I. Gibbs, Hyatt M. II. Khitrova, G. (Galina), 1959– . III. Peyghambarian, Nasser, 1954– . IV. Series. TA1522.N66   1990 621.36–dc20   90-9535

2154/3150(3011)-543210 – Printed on acid-free paper

Preface

This book is a progress report of research into practical applications of *nonlinear photonics* (activities incorporating nonlinear optical devices which use light instead of electronics to generate, communicate, process or analyze information). Currently, photonics uses only one nonlinear optical device, the diode laser, and it is used only as a linear light source. Photonic systems such as optical disc players, supermarket scanners, or the worldwide telecommunications system contain no other nonlinear optical devices. Modulation, switching, routing, decision-making and detecting are all being done with electronics and linear optoelectronic devices. This situation may soon change as nonlinear optical devices such as picosecond samplers and switches complement optoelectronic devices and help continue the rapid expansion of photonic capabilities and applications.

Chapter 1 takes a critical look at the problems impeding the progress of nonlinear photonics. The remainder of the book is descriptive, reporting the current status of research on nonlinear photonics: semiconductor nonlinear optical materials (Chap. 2), linear holographic optical interconnects (Chap. 3), proof-of-principle demonstrations of rudimentary nonlinear photonic systems using nonlinear interference filters (Chap. 4), the properties and applications of photorefractive crystals for image processing (Chap. 5), and nonlinear waveguide devices for switching and routing systems (Chap. 6). Authors succinctly summarize the past accomplishments and point to some hopes for the future, making this an ideal book for a new or seasoned researcher wanting to design and perfect nonlinear optical devices and to identify exciting applications in photonic systems.

Tucson, AZ *Hyatt M. Gibbs · Nasser Peyghambarian*
October 1989 *Galina Khitrova*

Contents

Index of Contributors

Gibbs, Hyatt M.
Optical Sciences Center, University of Arizona,
Tucson, AZ 85721, USA

Goodman, Joseph W.
Information Systems Laboratory, Department of Electrical Engineering,
Stanford University, Stanford, CA 94305, USA

Günter, Peter
Institut für Quantenelektronik, ETH-Hönggerberg,
CH-8093 Zürich, Switzerland

Hesselink, L.
Department of Aeronautics and Astronautics, Durand Building,
Stanford University, Stanford, CA 94305, USA

Huignard, Jean-Pierre
Thomson-CSF, Laboratoire Central de Recherches,
Domaine de Corbeville, F-91404 Orsay, France

Khitrova, Galina
Optical Sciences Center, University of Arizona,
Tucson, AZ 85721, USA

Koch, Stephan W.
Optical Sciences Center, University of Arizona,
Tucson, AZ 85721, USA

Kostuk, Raymond K.
Electrical and Computer Engineering, University of Arizona,
Tucson, AZ 85721, USA

Peyghambarian, Nasser
Optical Sciences Center, University of Arizona,
Tucson, AZ 85721, USA

Rajbenbach, Henri
Thomson-CSF, Laboratoire Central de Recherches,
Domaine de Corbeville, F-91404 Orsay, France

Silberberg, Yaron
Bellcore,
331 Newman Springs Road, Red Bank, NJ 07701-7040, USA

Smith, Peter W.
Bellcore,
331 Newman Springs Road, Red Bank, NJ 07701-7040, USA

Smith, S. Desmond
Department of Physics, Heriot-Watt University,
Edinburgh EH14 4AS, UK

Walker, Andrew C.
Department of Physics, Heriot-Watt University,
Edinburgh EH 14 4AS, UK

Wherrett, Brian S.
Department of Physics, Heriot-Watt University,
Edinburgh EH14 4AS, UK

1. Nonlinear Photonics: Prospects and Problems

H.M. Gibbs and G. Khitrova

Much progress has been made in the last 15 years in characterizing known materials, creating new materials, inventing new devices, and demonstrating logic and switching functions [1.1–17]. While the other five chapters of this book summarize that steady progress, it is also meaningful to ask why progress has been so slow. Why are so few nonlinear devices finding their way into photonic systems?

1.1 Nonlinear Materials

Since all nonlinear devices depend on nonlinear absorption and/or refraction, considerable effort has been devoted to the growth and characterization of known materials and the design of new materials.

For logic devices, much of the research has centered on *semiconductors*. The many-body theory of bulk-semiconductor band-edge optical nonlinearities is now well developed, quite successful in explaining experimental measurements, and extensively used in device modeling (Chap. 2) [1.14]. A similar formalism will soon be available for multiple-quantum-well inorganic semiconductors. In order to reduce the power of steady-state devices or the energy for pulsed devices, it would be highly desirable to enhance these already huge resonance nonlinearities even further. Neither the theory nor measurements of bulk and multiple-quantum-well crystals indicate how that could happen from inorganic semiconductors. Maybe some clever use of coupled wells or internal fields will negate that assessment. Further quantum confinement, such as quantum wires or dots, would increase the nonlinearity per unit volume but not the index change per absorbed photon [1.13]. Organic materials have been highly touted but have yielded no device competitive with inorganic semiconductors [1.18].

The heart of practically every proposed optical implementation of a neural network or real-time associative memory is a *photorefractive crystal* (Chap. 5). Admittedly, such crystals exhibit many attractive features, such as low-intensity (mW/cm^2) operation, same-frequency two-beam energy transfer, and self-pumped phase conjugation, that make them ideal show-business materials [1.8, 16, 17]. Nonetheless, there are numerous problems in using them for practical applications, including difficulty in growing large homogeneous crystals, fanning (am-

Springer Series in Electronics and Photonics, Vol. 30
Nonlinear Photonics   Editors: H.M. Gibbs · G. Khitrova · N. Peyghambarian
© Springer-Verlag Berlin Heidelberg 1990

plification of scattering), limitation on the number of images that can be stored in a single crystal, slow response time, instabilities, angular dependence of energy transfer (you cannot teach the crystal if a grating cannot be written), and freezing in the "learned" gratings.

Although a miracle material could still be discovered, there seems to be no fundamental-physics argument to predict it. Consequently, whereas the study of nonlinear materials should certainly be vigorously pursued, nonlinear photonics has moved strongly into the device phase. We must face the likelihood that nonlinear materials may not improve by an order of magnitude and must proceed with the optimization of devices based on existing optical nonlinearities.

1.2 Nonlinear Devices

Much progress has been made on nonlinear devices in which the light travels perpendicular to a nonlinear thin film (Chaps. 2 and 4). Most of these all-optical devices are nonlinear etalons with reflectors on both sides of the thin film. The achievement of picosecond decision-making [1.19], 50-ps recovery [1.20], 0.6-pJ switching [1.21], and arrays [1.22] larger than 100×100 has encouraged excessive speculation on potential applications to digital and neural optical computing.

The most thoroughly analyzed device is the *nonlinear etalon* operating as a single-wavelength bistable element or as a logic gate with inputs and outputs at different wavelengths [1.1]. Typical bistability holding power is 10 mW per device; typical energy per logic operation per device is 10 pJ. To utilize the "massive parallelism of optics" would then require, assuming 10^6 elements per plane, a formidable 10 kW/plane cw for bistability; or 10 kW/plane average for a 1-GHz logic cycle rate and 10^{15} bits/s. How can the power and energy be reduced if the nonlinearities are not likely to be enhanced substantially? One way is to *reduce the size* of the device. Microresonators of GaAs multiple quantum wells with a nominal diameter of $2 \mu m$ have been operated as NOR gates with as little as 0.6 pJ as a result of the reduced size and waveguiding [1.21]. However, such miniaturization leads to formidable problems in focusing onto every one of a large array of pixels and in re-imaging to the next array. Some of the proposals to interconnect such microresonators in effect form linear waveguides between them; this approach has many of the restraints of planar waveguides. Engineering compromises may result in slightly larger devices demanding higher switching energies. *Nonlinear waveguides* [1.23] do not seem to be the solution either. The usual guided-wave argument, "reduce the power by increasing the interaction length," works for optical fibers because they are almost transparent and the nonlinearity is nonresonant and hence relatively weak. But to reduce the energy of logic devices, a resonant nonlinearity must be used. Then to reduce absorption, a larger detuning is necessitated by the longer length of a waveguide device, resulting in a reduced index change per absorbed photon. Power problems may be reduced using gain devices [1.24–26], but diode-laser-length devices are very sensitive to temperature and input wavelength, and massive parallelism using devices with a cross section of $0.2 \times 1.0\,cm^2$ seems hopeless. *So we conclude that there is no reason to expect orders-of-magnitude reduction in power and energy requirements.*

Another unsolved problem is the *mode of operation.* Optical-bistability and NOR-gate operations are the most often discussed [1.1], and both have serious problems. *Optical bistability* is a single-wavelength operation: the output and inputs have the same wavelength, facilitating the cascading of devices which is essential for digital optical computing (Sects. 2.3.1 and 4.2). As discussed in Chap. 4, bistability has been used for a number of low-speed computing demonstrations. While some believe that the parallelism of optics will permit special-purpose digital optical computers to compete with electronic computers, we have serious reservations. *The problem in applying optical bistability to high-speed computing is that the differential gain decreases as the pulse length approaches the medium response time* (Sect. 3.7) [1.27]. To obtain a gain of 2.5 in a typical GaAs etalon requires a pulse length 20 times longer than the medium response time. The latter is typically 3–30 ns, so that pulse lengths of 0.1–1 μs are usually used to observe bistability. For operation at a 1-ns cycle time, the recovery time should be on the order of 10 ps; this is probably achievable, but would cause the power requirement to increase as much as the carrier lifetime is decreased. These considerations, and others based on device and input tolerances, have led us to *abandon bistability for high-speed decision-making.*

A more promising mode of operation is the *NOR-gate* (Sect. 2.4.5). With NOR-gate operation, the inputs are at a wavelength for complete absorption, and the probe beam (whose transmission is the output) is at a different wavelength having low absorption. This mode of operation is very efficient, has a gain of 5 or 10, and yields the answer to a logic question in a few picoseconds. The fact that the device cannot be used again until the medium has recovered is no problem in many cases, since the cycle time can be made longer than the medium response time. *The problem in using the NOR gate for digital computing is cascading: the output has a different wavelength from the input.* One solution is to shift the band edge and Fabry-Perot peak of the next device; for example, by increasing the well thickness of GaAs/AlGaAs multiple quantum wells. Clearly, one can do this for only a few stages (5 or 10). Another approach to solving the cascading problem would be to achieve genuine two-wavelength operation by using opposite sides of two isolated absorption peaks so that the absorption peak of every other device would be different; then the output of one device would be the ideal input for the subsequent device. Possibilities would be the exciton peaks at low temperature in bulk GaAs and in AlGaAs with low Al concentration or two nearby transitions of two isotopes of an alkali atom. Neither material is attractive for a practical device. *Although cascading through a few stages could be done, an infinitely cascadable mode of operation for high-speed, all-optical decision-making has not been found.*

3

1.3 Applications of Nonlinear Photonics

Nonlinear photonic systems can be divided into digital optical computing, optical neural computing, and photonic switching. Although system applications of nonlinear optical devices seem premature given the many problems discussed above, early attempts at prototype sytems (Chaps. 4 and 5) have been constructive in clarifying device shortcomings.

The impressive success of digital electronic computers and the problem with nonlinear optical devices (cascading, energy, and tolerances) make us *doubt that nonlinear photonics will have an impact on computing*. If it does, it will probably be as an ultrafast preprocessor or as an ultrafast switch that removes some bottleneck. We are also *skeptical about nonlinear photonics contributing to optical neural computing*. Some system designs include nonlinear etalons for decisions-making or for controlling sigmoidal transmission to achieve backward error propagation. Since GaAs etalons and ZnSe interference filters require $\geq 1\,\mathrm{kW/cm^2}$, they are incompatible with efficient photorefractive crystals requiring $\simeq 1\,\mathrm{mW/cm^2}$, unless one is performing whole-image decision-making (Sect. 4.4.4e). Unless slow, low-power, all-optical devices can be developed, linear electro-optic devices such as the SEED (self electro-optic effect device) may be the only practical solution. Such electro-optic devices may be able to preserve the phase as is often required. However, no photorefractive crystal works well at the GaAs SEED wavelength. In fact, with all the problems listed before, photorefractive crystals may be the bottleneck in implementing neural networks optically. In addition, in the neural net area, as in the digital computing area, the electronic competition is quite formidable and is expanding rapidly. Perhaps optics should be used only for the vitally important interconnect function (Chap. 3).

Finally, nonlinear optics can be applied to *photonic switching* for multiplexing, time re-ordering, high-speed routing, sampling, etc. (Sects. 2.3.5–7, 4.4.6 and Chap. 6) and for image processing, dynamic interconnects, etc. (Chap. 3). After all, optics leads electronics by far in short-pulse generation, transmission, and diagnostics [1.9, 10]. Several switches that open or close in a picosecond have been demonstrated, with recovery times varying from 1 ps to several nanoseconds. Multiple quantum wells [1.28, 29] (even though dismissed in Chap. 6) and optical glass fibers [1.30] have been made into picosecond switches, and many organic materials exhibit the appropriate response times [1.18]. So switching seems to be the area of nonlinear photonics where nonlinear optics is making unique contributions; i.e., accomplishing feats not performed at all or as well by electronics. Clearly, the disadvantages of high switching energy and two-wavelength operation are not nearly as serious for a single switch as for an optical computer. Also in this area, one or a few devices can be the essential element that makes possible a faster overall system, *making photonic switching the area most ripe for nonlinear photonic expansion*.

1.4 Conclusion

We believe that most current research efforts in nonlinear photonics should still be directed toward nonlinear materials and devices. While system awareness is crucial so that relevant devices are designed and tested, until the basic building blocks are constructed, extensive design of nonlinear photonic systems is premature. We hope that this book, with its summary of the present situation, will support careful material and device research which can then facilitate the design and construction of nonlinear photonic systems. Once we physicists and device engineers make sufficient improvements in nonlinear optical devices, systems engineers will recognize their usefulness and incorporate them.

References

1.1   H.M. Gibbs: *Optical Bistability: Controlling Light with Light* (Academic, New York 1985)
1.2   C.M. Bowden, M. Ciftan, H.R. Robl (eds.): *Optical Bistability* (Plenum, New York 1981)
1.3   C.M. Bowden, H.M. Gibbs, S.L. McCall (eds.): *Optical Bistability 2* (Plenum, New York 1984)
1.4   H.M. Gibbs, P. Mandel, N. Peyghambarian, D. Smith (eds.): *Optical Bistability III*, Springer Proc. Phys., Vol. 8 (Springer, Berlin, Heidelberg 1986)
1.5   W. Firth, N. Peyghambarian, A. Tallet (eds.): *Optical Bistability IV*, J. de Phys. **49**, Colloq. C2, suppl. (1988)
1.6   B.S. Wherrett, S.D. Smith (eds.): *Optical Bistability, Dynamical Nonlinearity and Photonic Logic* (Royal Society, London 1985)
1.7   H.H. Szu, R.F. Potter (eds.): *Optical and Hybrid Computing*, SPIE, Vol. 634 (SPIE, Bellingham, WA 1987)
1.8   R.A. Fisher: *Optical Phase Conjugation* (Academic, New York 1983)
1.9   F.J. Leonberger, C.H. Lee, F. Capasso, H. Morkoç (eds.): *Picosecond Electronics and Optoelectronics II*, Springer Ser. Electron. Photon., Vol. 24 (Springer, Berlin, Heidelberg 1987)
1.10   T.K. Gustafson, P.W. Smith (eds.): *Photonic Switching*, Springer Ser. Electron. Photon., Vol. 25 (Springer, Berlin, Heidelberg 1988)
1.11   T. Kobayashi (ed.): *Nonlinear Optics of Organics and Semiconductors* Springer Proc. Phys., Vol. 36 (Springer, Berlin, Heidelberg 1989)
1.12   N. Peyghambarian (ed.): *Optical Computing and Nonlinear Materials*, SPIE, Vol. 881 (SPIE, Bellingham, WA 1988)
1.13   H. Haug, L. Banyai (eds.): *Optical Switching in Low-Dimensional Systems* (Plenum, New York 1989)
1.14   H. Haug (ed.): *Optical Nonlinearities and Instabilities in Semiconductors* (Academic, New York 1989)
1.15   A.N. Chester, S. Martellucci (eds.): *Nonlinear Optics and Optical Computing* (Academic, Cambridge 1988)
1.16   P. Günter, J.P. Huignard (eds.): *Photorefractive Materials and Their Applications I: Fundamental Phenomena*, Topics Appl. Phys., Vol. 61 (Springer, Berlin, Heidelberg 1988) and *Photorefractive Materials and Their Applications II: Survey of Applications*, Topics Appl. Phys., Vol. 62 (Springer, Berlin, Heidelberg 1989)
1.17   Optical Computing Topical Meetings (Optical Society of America, Washington, DC 1985, 1987, 1989)
1.18   D.S. Chemla, J. Zyss (eds.): *Nonlinear Optical Properties of Organic Molecules and Crystals* (Academic, New York 1987)
1.19   A. Migus, A. Antonetti, D. Hulin, A. Mysyrowicz, H.M. Gibbs, N. Peyghambarian, J.L. Jewell: Appl. Phys. Lett **46**, 70 (1984)

1.20 Y.H. Lee, H.M. Gibbs, J.L. Jewell, J.F. Duffy, A.C. Gossard, W. Wiegmann, J.H. English, T. Venkatesan: Appl. Phys. Lett **49**, 486 (1986)

1.21 J.L. Jewell, A. Scherer, S.L. McCall, A.C. Gossard, J.H. English: Appl. Phys. Lett. **51**, 94 (1987)

1.22 T. Venkatesan, B. Wilkens, Y.H. Lee, M. Warren, G.R. Olbright, H.M. Gibbs, N. Peyghambarian, J.S. Smith, A. Yariv: Appl. Phys. Lett. **48**, 145 (1986)

1.23 G.I. Stegeman, E.M. Wright, N. Finlayson, R. Zanoni, C.T. Seaton: J. Lightwave Tech. **6**, 953 (1988)

1.24 G.J. Lasher: Solid-State Electron. **7**, 707 (1964)

1.25 Ch. Harder, K.Y. Lau, A. Yariv: Appl. Phys. Lett. **39**, 382 (1981); ibid. **40**, 124 (1982)

1.26 H. Kawaguchi: Electron. Lett. **17**, 741 (1981); Appl. Phys. Lett. **41**, 702 (1982);
W.F. Sharfin, M. Dagenais: Appl. Phys. Lett. **46**, 819 (1985)

1.27 R. Jin, C. Hanson, M. Warren, D. Richardson, H.M. Gibbs, N. Peyghambarian, G. Khitrova, S.W. Koch: Appl. Phys. B **46**, 61 (1988)

1.28 P. Li Kam Wa, P.N. Robson, J.P.R. David, G. Hill, P. Mistry, M.A. Pate, J.S. Roberts: Electron. Lett. **22**,1129 (1986)

1.29 R. Jin, C.L. Chuang, H.M. Gibbs, S.W. Koch, J.N. Polky, G.A. Pubanz: Appl. Phys. Lett. **53**, 1791 (1988)

1.30 S.R. Friberg, Y. Silberberg, M.K. Oliver, M.J. Andrejco, M.A. Saifi, P.W. Smith: Appl. Phys. Lett. **51**, 1135 (1987)

2. Semiconductor Nonlinear Materials

N. Peyghambarian and S.W. Koch

With 33 Figures

A large variety of optical nonlinearities in laser-excited semiconductors has been observed in the last few years (for recent reviews see, e.g. [2.1–8]). The magnitude of these nonlinearities can become quite large because of the resonance enhancement in the spectral vicinity of the fundamental absorption edge.

In direct-gap semiconductors, the absorption of photons excites electrons from the valence band into the conduction band. As electrically charged quasiparticles, conduction-band electrons and valence-band holes interact through the Coulomb potential. This interaction is repulsive for equally charged quasiparticles, i.e. within the same band, and it is attractive for oppositely charged quasiparticles in different bands. The attractive interband interaction causes a strong correlation between electrons and holes and leads to the formation of bound states or excitons. In the so-called Wannier approximation, excitons may be regarded as hydrogen-atom-like quasiparticles, which are characterized by a Bohr radius, binding energy, etc. (for more details see [2.9, 10]).

Laser excitation of semiconductor materials like GaAs, CdS, CdSe, or ZnS creates a high density of electrons and holes. These quasiparticles form a quantum mechanical system in which nonlinearities arise through many-body effects such as screening of the Coulomb potential, reduction of the band-gap, filling of the bands and of the states. In most situations, for not too fast excitation pulses, the changes of the semiconductor absorption and dispersion depend on the density of excited electron–hole pairs and therefore on the intensity of the exciting laser light.

The dominant contribution to the nonlinearity depends on the type of semiconductor and the band-structure parameters. For example, at room temperature, the dominant nonlinear mechanism in high quality ZnSe is exciton screening and broadening [2.11], while in GaAs it is the band filling and screening of the continuum states [2.12]. Generally, for semiconductors with small band-gap energies, such as InSb, InAs, and CdHgTe, the exciton is only weakly bound and spectrally broad so that it cannot be resolved in the absorption spectra (see Table 2.1 for comparison of the various material parameters). Therefore, band filling and screening of the continuum enhancement provide the main nonlinearity. On the other hand, the excition binding energy for wide-gap bulk semiconductors is large and the exciton Bohr radius is small. Consequently, at least for low-temperature conditions, excition screening provides a major contribution to the nonlinear index of refraction.

Springer Series in Electronics and Photonics, Vol. 30
Nonlinear Photonics   Editors: H.M. Gibbs · G. Khitrova · N. Peyghambarian
© Springer-Verlag Berlin Heidelberg 1990

Table 2.1. Important parameters of several semiconductors. The values for the band-gap energy E_g and broadening parameter Γ correspond to $T = 300\,\mathrm{K}$, unless otherwise specified. E_R is the exciton binding energy, m_e and m_h are the effective masses of the electron and hole, respectively. Γ_0 and Γ_1 are the intensity-independent and intensity-dependent broadening parameters, respectively. An average value for the hole effective mass (optical mass) is given, i.e. $m_h = (m_{h_1}^2\, m_{h_{11}})^{1/3}$. (See [2.13] for additional material parameters)

	E_g [eV]	E_R [meV]	$m_e m_0$	$m_h m_0$	$\Gamma_0 E_R$	$\Gamma_1 E_R$ [cm^3]
GaAs	1.420	4.2	0.0665	0.52	1.25	2×10^{-18}
ZnSe	2.735	18	0.16	4.75	1.	2.5×10^{-18}
CdS	2.520	27	0.235	1.35	1.	3×10^{-19}
CuCl	3.4 at 4.2 K	190	0.41	3.6	–	–
InSb	0.23 at 77 K	0.5	0.0145	0.4	2	–
InAs	0.345	–	0.024	0.33	–	–
Ga$_{0.47}$ In$_{0.53}$As	0.75	2.7	0.041	0.4	–	–
CdSe	1.74	15.25	0.13	0.5	–	–
ZnTe	2.3	11	0.096	0.6	–	–
CdTe	1.45	9	0.1	0.4	–	–

Such optical nonlinearities may be used to demonstrate optical switches, gates and bistable elements under suitable conditions. For most of these devices, the nonlinearity of the material is combined with a feedback mechanism for their operation. The most common device consists of a nonlinear medium between two partially reflecting mirrors, forming a nonlinear etalon. Depending on whether the feedback is provided electronically or optically, the bistability is hybrid [2.14, 15] or all-optical [2.5], respectively [2.4]. In usual dispersive or absorptive optical bistability, the feedback is external (e.g., external Fabry-Perot mirrors). Induced absorption or increasing absorption bistability occurs when the feedback is intrinsic (or internal) [2.16–28].

Optical bistable devices may perform gating operations under different operating conditions. More than one input beam is employed for optical gating. For example, to demonstrate a NOR gate, the transmission of a probe pulse is monitored in the presence or absence of two input pulses. The probe transmission, which is the gate output, is adjusted (in the absence of any inputs) to the "high" state by matching its peak wavelength to one of the Fabry-Perot peaks. The input pulse changes the index of refraction of the material, shifts the Fabry-Perot peak away from the probe wavelength, and thereby forces the gate output to the "low" state. The second input pulse shifts the Fabry-Perot peak even further and the probe transmission stays in the "low" state. Various gating operations such as AND, OR, and NOR have been demonstrated in bulk GaAs and GaAs/AlGaAs MQWs [2.5].

Additional to the optical nonlinearities in bulk semiconductors, sophisticated semiconductor microstructures are being investigated which provide quantum confinement of the excited elementary excitations. Examples are the quasi-two-dimensional semiconductor quantum wells, the quasi-one-dimensional quantum wires, and the quasi-zero-dimensional quantum dots. Experimental realization of these semiconductor nanostructures is very difficult because of the stringent requirement on the size distribution. At present, most of the effort is concentrated on II-VI and III-V compounds. Glasses doped with CdS and CdSe microcrystallites represent one area of research [2.29–33]. Various etching techniques in GaAs-based samples provide another area for quantum wires and boxes [2.34–36].

At present, the theory of quantum-confined structures is much more advanced than the experiments. Several theoretical groups have treated various regimes of quantum confinement [2.37–43].

2.1 Optical Nonlinearities in Bulk Semiconductors

In the spectral vicinity of the fundamental absorption edge, the electronic optical nonlinearities of semiconductors are mainly determined by the interaction processes among the laser-excited electrons and holes. Before we describe the theoretical analysis of these interactions, we present simple arguments to familiarize the nonexpert with the subject. This is done with figures for representative materials and a discussion of the basic nonlinear effects.

Figure 2.1a shows the computed absorption spectrum of low-temperature GaAs. This spectrum is representative of a medium-gap bulk semiconductor near the band edge. The main features of the spectrum at low carrier density (curve 1) are the narrow excitonic lines (only $n = 1$ and $n = 2$ fundamental lines are shown) on the low-energy side of the absorption edge. The energies are plotted relative to the band-gap; thus the (unrenormalized) absorption edge is located at position 0.0 on the horizontal axis. At frequencies higher than the excitonic lines is the broad band of continuum states. These states are affected by Coulomb enhancement, which

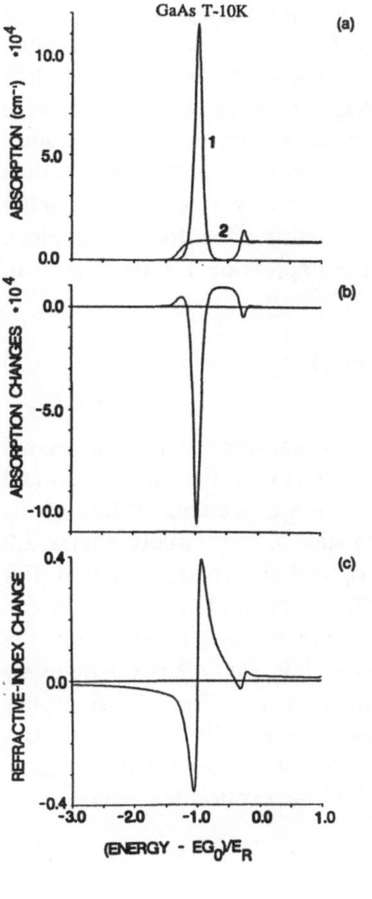

Fig. 2.1a–c. Excitonic optical nonlinearity in semiconductors. (a) The absorption spectra of GaAs at 10 K at low (curve 1) and high (curve 2) excitation intensities. (b) The change in absorption coefficient, $\Delta\alpha$. (c) The change in index of refraction which is obtained by a Kramers-Kronig transformation of $\Delta\alpha$.

9

changes the hypothetical square-root energy dependence of the free-particle absorption to the realistic semiconductor absorption with an almost constant energy dependence. When additional free electron-hole pairs are generated (for example by optical excitation), the exciton ionizes as a result of the screening of the Coulomb potential by additional carriers. Curve 2 in Fig. 2.1a exhibits the modified absorption spectrum at a larger carrier density where the excitonic resonances have disappeared. The change in absorption coefficient, $\Delta\alpha = \alpha(N) - \alpha_0$, can be obtained by direct subtraction of curves one and two in Fig. 2.1a. The result is plotted in Fig. 2.1b. A Kramers-Kronig transformation of $\Delta\alpha$ gives the nonlinear change in the index of refraction Δn as plotted in Fig. 2.1c:

$$\Delta n(\omega) = -\frac{\hbar c}{\pi} P \int_{\omega_1}^{\omega_2} \frac{\Delta\alpha(\omega')}{\omega'^2 - \omega^2} d\omega' , \tag{2.1}$$

where P stands for the principal value of the integral. It is noted that Δn is negative on the low-energy side of the exciton and positive on the high-energy side. A negative index change corresponds to a self-defocusing optical nonlinearity.

The optical nonlinearity resulting from bleaching of excitonic resonances as just described usually has a large magnitude and a relatively rapid response time. This nonlinearity is often dominant, mainly at low temperatures, in the spectral vicinity of the excition, and therefore it can be used in a narrow frequency region. Another consequence of the Coulomb interaction among the laser-excited electronic excitations is the effect of band-gap shrinkage (band-gap renormalization). The energy gap shifts to lower energies with increasing carrier density because of the screening and intraband exchange and correlation effects. The carriers within each band reduce the total energy of the system by avoiding each other (exchange effect for particles with equal spin, correlation effect for all particles). In the screened Hartree-Fock approximation, the expression for the band-gap reduction is [2.7, 44–46]

$$\delta E_g = \sum_{q \neq 0} [V_s(q) - V(q)] - \sum_{q \neq 0} V_s(q)(n_{e,q} + n_{h,q}) \tag{2.2}$$

where the first term is often called the Coulomb-hole self-energy, and the second term is the screened exchange. In (2.2) $V(a)$ and $V_s(q)$ are the unscreened and screened Coulomb potential, and $n_{e,q}$ and $n_{h,q}$ are the expectation values of the electron and hole density operators at momentum state q, respectively. Figure 2.2 shows the effect of band-gap shrinkage on the optical absorption spectrum. For the example of the typical II-VI compounds CdS, we have plotted the semiconductor absorption at low excitation densities (linear spectrum I) together with the spectrum under higher excitation (curve 2). Curve 1 in Fig. 2.2 is qualitatively very similar to curve 1 in Fig. 2.1a. However, in curve 2 of Fig. 2.2, we clearly see that the absorption has increased in the regime spectrally below the exciton resonance. This increase is a direct consequence of the band-gap reduction, which in this case has shifted the band edge below the exciton resonance.

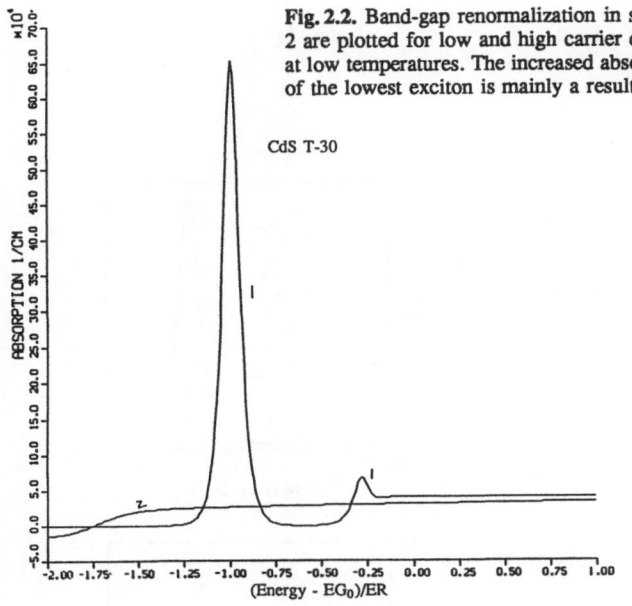

Fig. 2.2. Band-gap renormalization in semiconductors. Curves 1 and 2 are plotted for low and high carrier densities, respectively, in CdS at low temperatures. The increased absorption on the low energy side of the lowest exciton is mainly a result of band-gap renormalization

In narrow-gap semiconductors where the exciton is only weakly bound, band-filling effects may become the dominant source of the optical nonlinearity. The band-filling effect is the result of the Pauli exclusion principle. The band states that are filled by electrons and holes are no longer available for absorption due to the Pauli principle. Because of the rapid thermalization process in the system of electronic excitations, electrons and holes are most likely to fill the states close to the extrema of their respective bands. Therefore, the semiconductor becomes more and more transparent for energies slightly above the band-gap and it appears as if the onset of absorption shifts to higher energies. This effect is the dominant source of noninearities in InSb, for example, as shown in Fig. 2.3a. Again, curve 1 is plotted for low carrier density and curve 2 for a higher carrier density. The absence of an excitonic feature in this low-temperature narrow-gap material is clear in this figure. The change in absorption coefficient, $\Delta\alpha$, is shown in Fig. 2.3b and its Kramers-Kronig transformation, Δn, is in Fig. 2.3c. The nonlinear index is negative at frequencies below the absorption edge, similar in sign to the exciton bleaching case.

In wide-gap semiconductors such as CuCl, CuBr, and HgI$_2$, the exciton binding energy E_b is relatively large. For example, E_b is $\sim 190\,\text{meV}$ in CuCl. In these materials, the bound state of two excitons, which is called an excitonic molecule or a biexciton, is also stable. The biexciton energy is twice the exciton energy minus the binding energy of the biexciton ($E_{xx} = 2E_x - E_{bxx}$, where E_{bxx} is the biexciton binding energy). Because the biexciton complex has two electrons and two holes, it can be excited directly by two-photon absorption:

$$2\hbar\omega = E_{xx} \qquad \text{or} \qquad \hbar\omega = E_x - \frac{E_{bxx}}{2} . \qquad (2.3)$$

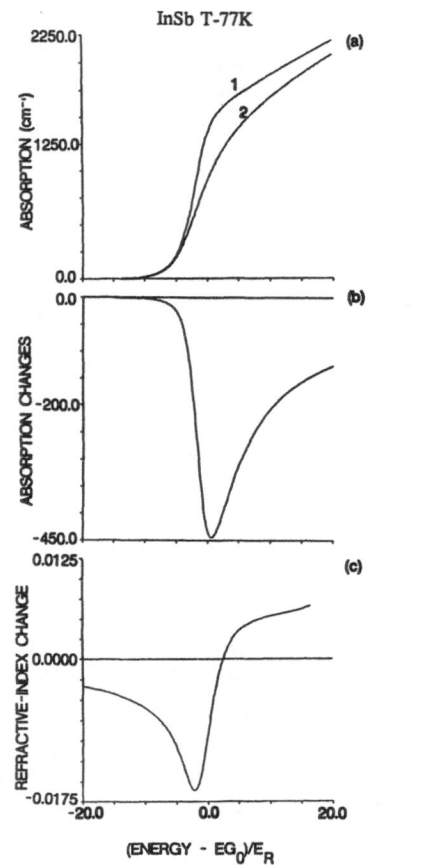

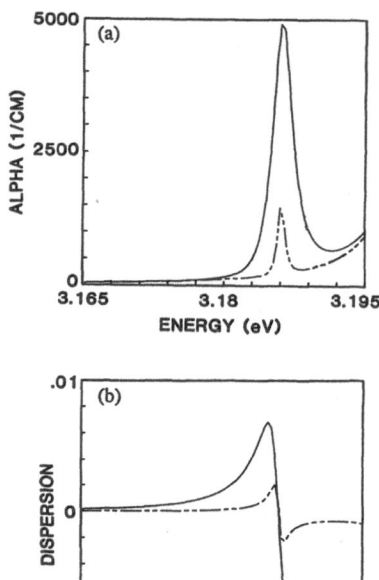

Fig. 2.3a–c. Band-filling nonlinearity in narrow-band-gap semiconductors. (a) The absorption spectra of InSb at low temperatures for low (curve 1) and high (curve 2) carrier densities. (b) The change in absorption coefficient, $\Delta\alpha$. (c) The change in index of refraction, Δn

Fig. 2.4a,b. Biexciton optical nonlinearity in wide-gap semiconductors. (a) The absorption spectra for low, $n_p = 5 \times 10^{15}\,\text{cm}^{-3}$ (dashed curve) and high, $n_p = 5 \times 10^{16}\,\text{cm}^{-3}$ (solid curve) polariton densities. The polariton density of $5 \times 10^{15}\,\text{cm}^{-3}$ corresponds to $\cong 1\,\text{MW/cm}^2$ intensity at biexciton resonance. (b) The change in index of refraction Δn for the same parameters as Fig. 2.4a

Therefore, the energy of the photon needed to create the biexciton is below the exciton energy by half of the biexciton binding energy. The biexciton resonance is then usually detected below the exciton. Biexcitons are formed at higher laser excitations in wide gap semiconductors.

Figure 2.4a shows the calculated biexciton absorption in CuCL [2.47]. The absorption peak increases at higher laser intensities (or higher polariton densities) as shown by the solid curve. The laser-induced change in the absorption coefficient produces the corresponding change in the refractive index that is exhibited in Fig. 2.4b. Here, the nonlinearity below the biexciton has a positive sign in contrast to the exciton bleaching and band filling.

Finally, the different semiconductor materials experience a temperature change as photons are absorbed. In passive semiconductors, the vast majority (> 90%) of all laser-excited electron–hole pairs recombine nonradiatively, giving their energy to the crystal through phonon excitation, which leads to lattice heating. In most materials, the temperature change causes a shift of the absorption spectrum to lower frequencies, but there are also some exceptions such as CuCl where the bandgap widens as it is heated. The low-energy shift of the absorption spectrum gives rise to a nonlinear index change, which is schematically shown in Fig. 2.5. As can be seen from this figure, a thermally induced index change is positive below the band edge. This property of a thermal nonlinearity, together with its slow response time, helps to distinguish it from optical nonlinearities of electronic origin. Due to the ease in obtaining thermal nonlinearities, and their existence at room temperature, researchers have used them extensively. However, for practical applications thermal nonlinearities are not desirable.

The above discussion on nonlinear mechanisms is mostly descriptive and schematic. In more rigorous theoretical treatments, the interactions between the laser-excited electrons and holes may be incorporated into a dielectric response function which depends parametrically on the pair density N. Many theoretical treatments of the semiconductor nonlinearities use quantum mechanical Green's functions [2.48–52]. The results have been applied successfully to analyze various semiconductor nonlinearities.

Unfortunately, however, the Green's function results are often difficult to use for studies of optical bistability, optical logic gating, and other semiconductor instabilities, since in most cases one has to perform prohibitively extensive numerical solutions of integral equations (Bethe-Salpeter equation). In the present review, we therefore concentrate on a less sophisticated, partly phenomenological description of the optical nonlinearities in an electron–hole plasma, which includes the most important nonlinearities and which has the advantage that it can be easily evaluated [2.7, 53]. For a recent review of the equilibrium and nonequilibrium Green's functions theory, we refer to the article by *Haug* [2.52].

In most theoretical descriptions, one uses the concept that the optical response of a laser ex-

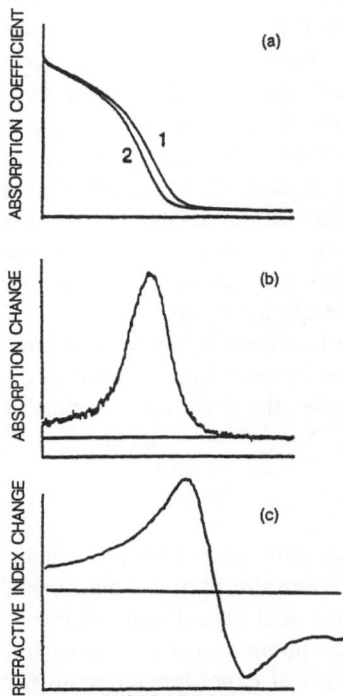

Fig. 2.5a–c. Thermal optical nonlinearity in semiconductors. (a) Schematics of the absorption spectra in the vicinity of the band-gap for low (curve 2) and high laser intensities (curve 1). (b) The change in absorption coefficient. (c) The change in the index of refraction

13

cited semiconductor can be computed in the scheme of linear response theory, but the material parameters are renormalized and depend parametrically on the electron-hole-pair density. This approach is best justified under quasi-equilibrium conditions in the electron–hole system, which is typically established on the time scale of several hundred femtoseconds through the rapid collisions between the elementary excitations. In quasi equilibrium, the electrons and holes are then described by Fermi functions within their bands. Quasi chemical potentials are defined through the particle number conservation, and an effective plasma temperature is introduced. The Fermi functions regulate the occupation of the energy states in the bands. This band filling and the simultaneous screening of the Coulomb interaction are the basic microscopic reasons for the electronic optical nonlinearities of semiconductors.

In the plasma theory for semiconductor nonlinearities [2.53], the optical absorption spectrum is computed using a partly phenomenological generalization of the Elliott-formula [2.54] for the absorption coefficient $\alpha(\omega)$:

$$\alpha(\omega) = A \tanh\left[\frac{\beta}{2}(\hbar\omega - \mu)\right] \sum_\lambda |\phi_\lambda(r = 0)|^2 \delta_\Gamma(\hbar\omega - E_\lambda) . \qquad (2.4)$$

The prefactor A is proportional to the absolute square of the dipole matrix element between valence and conduction band, $\beta = 1/k_B T$, $\mu = \mu_e + \mu_h$ is the chemical potential of the electron–hole system, $\phi_\lambda(r)$ is the pair wavefunction, E_λ is the corresponding energy eigenvalue, and δ_Γ is a broadened δ function (line-shape function). The phenomenological linewidth is taken as $\Gamma = \Gamma_0 + \Gamma_1(N)$, where N is the density of electron–hole pairs. Both the electron-hole-pair wavefunction and the energy eigenvalue have to be computed from the Wannier equation with the screened interaction potential. The Wannier equation is the Schrödinger equation for the electron-hole relative motion with the potential energy given by the Coulomb potential. No analytic solutions exist in the general case, but one has a very good approximation if one neglects state-filling effects in the Wannier equation, and if one replaces the screened Coulomb potential by the Hulthén potential [2.55] for which the eigenfunctions and eigenvalues are known [2.56]. These effects are not too important for higher temperatures or for low temperatures when the electron-hole pair density is not too large.

In the following, we describe in more detail the nonlinear behavior of several bulk semiconductors. The plasma theory is evaluated for these materials and good agreement with experiment is obtained. In each case, the main contribution to the nonlinearity is identified.

2.1.1 GaAs

The nonlinear absorptive changes in laser-excited semiconductors are usually measured employing a pump-probe scheme. The spectrally narrow pump beam is tuned above the band-gap of the material to generate a high-density electron-hole plasma, and the broad-band probe monitors the pump-induced transmission changes. Figure 2.6 exhibits the schematic of a typical experimental setup for

STEADY-STATE EXPERIMENT

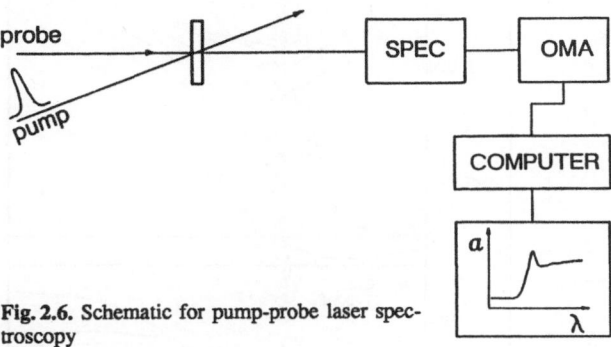

Fig. 2.6. Schematic for pump-probe laser spectroscopy

such a measurement. Two laser beams, which usually originate from the same laser source, pump two dye lasers. One dye laser has a narrow frequency output which is used as the pump beam. The second dye laser output is spectrally broad and is used as the probe beam. In this second dye laser the cavity is modified by simply inserting a blocking element to prevent the laser oscillation from reaching the end mirror so that the laser output is just the broad fluorescence of the dye. Both beams are focused on the same spot on the sample with the pump beam having a larger spot size to ensure a complete spatial overlap and to eliminate carrier diffusion problems. In the event that the two laser beams are pulses, a complete temporal overlap is also obtained. The transmission of the probe beam is detected using a spectrometer and an optical multichannel analyzer and the data are collected by a computer. The pump transmission is blocked from reaching the detector by both polarization and angle discrimination.

Applying this experimental technique to bulk GaAs at room temperature produced the result shown in Fig. 2.7 [2.12, 57, 58]. The top curve in Fig. 2.7a is the linear absorption spectrum showing a very weak exciton feature. The exciton is broadened by phonon interaction. Most of the negative absorption changes at small pump intensities come from the exciton bleaching. As the pump intensity increases, broad absorption changes in the band begin to take place and become dominant. Applying the Kramers-Kronig transformation to the measured absorption differences, gives the dispersive changes Δn shown in Fig. 2.7b. Even though this is an indirect method of measuring the nonlinear index, it has the advantage of providing correct values of Δn using a simple experimental setup. The validity of this technique has been confirmed by several direct interferometric measurements [2.59]. The maximum value of the refractive-index changes is approximately -0.06 at about 3 to 4 meV below the exciton energy. The nonlinear dispersive changes far below the band edge show nonresonant characteristics rather than resonant exciton behavior. The rising slopes at the high-energy end of Fig. 2.7b are artifacts arising from the finite integration limits (1.38–1.50 eV) set in the Kramers-Kronig transformation of the experimental data.

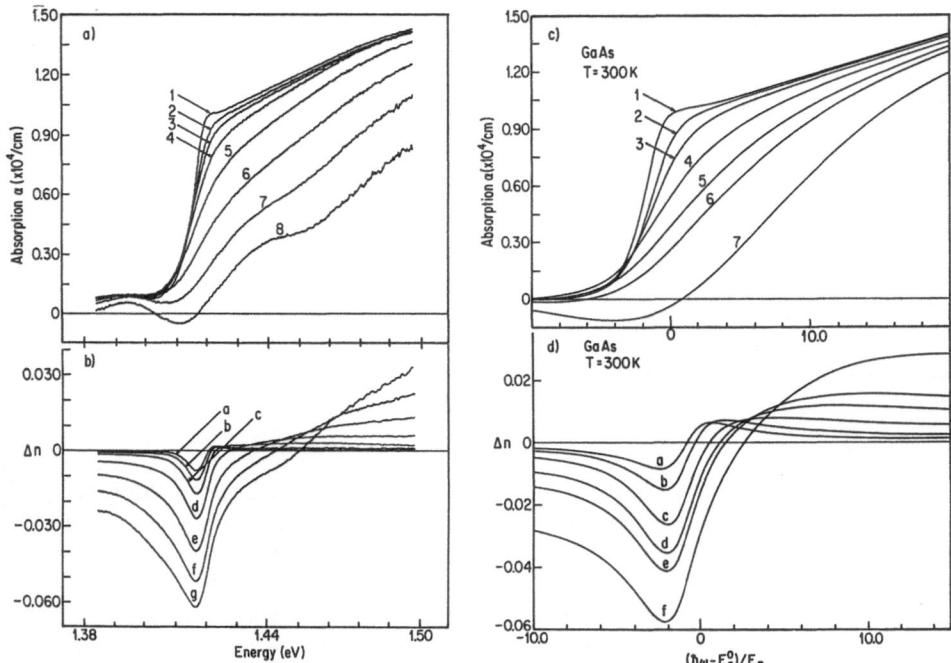

Fig. 2.7a–d. Room-temperature bulk GaAs optical nonlinearities: experiment and theory. (a) Experimental absorption spectra for different excitation powers P: (1) 0, (2) 0.2, (3) 0.5, (4) 1.3, (5) 3.2, (6) 8, (7) 20, (8) 50 mW on a 15-μm diameter spot. (b) Nonlinear refractive index changes corresponding to the measured absorption spectra. Curves a–g are obtained by the Kramers-Kronig transformation of the corresponding experimental data (2)–(8) in (a). (c) Calculated absorption spectra for different electron-hole pair densities N: (1) 10^{15} cm^{-3}, (2) 8×10^{16} cm^{-3}, (3) 2×10^{17} cm^{-3}, (4) 5×10^{17} cm^{-3}, (5) 8×10^{17} cm^{-3}, (6) 10^{18} cm^{-3}, (7) 1.5×10^{18} cm^{-3}. $E_g^0 = 1.420$ eV and $E_R = 4.2$ meV. (d) Calculated nonlinear refractive index changes. Curves a–f are obtained from curves (2)–(7) in (c), respectively

To apply the plasma theory to room-temperature bulk GaAs, we used the material parameters at $T = 300$ K, given in Table 2.1. Examples of the computed absorption spectra are shown in Fig. 2.7c [2.12, 57]. The low-density absorption curve 1 is the spectrum of the unexcited crystal. It exhibits a shoulder which is caused by the $1s$-exciton line broadened through phonon interactions. For higher plasma densities N, this shoulder vanishes because the increased carrier concentration causes strong screening, which weakens the attractive Coulomb potential between electron and hole. The exciton ionizes at the critical (Mott) density. This causes a decreasing absorption around the energetic position of the exciton line (changes from curve 1 in Fig. 2.7c to curve 2). However, the density-dependent effects due to band filling and screening of the continuum Coulomb enhancement are more important for room-temperature conditions. They are essentially responsible for the absorption changes in curves 3–7 in Fig. 2.7c. At the highest plasma density (curve 7), band filling causes a region of negative absorption, i.e. optical gain, in the spectral region between the renormalized gap and the quasi-chemical potential μ.

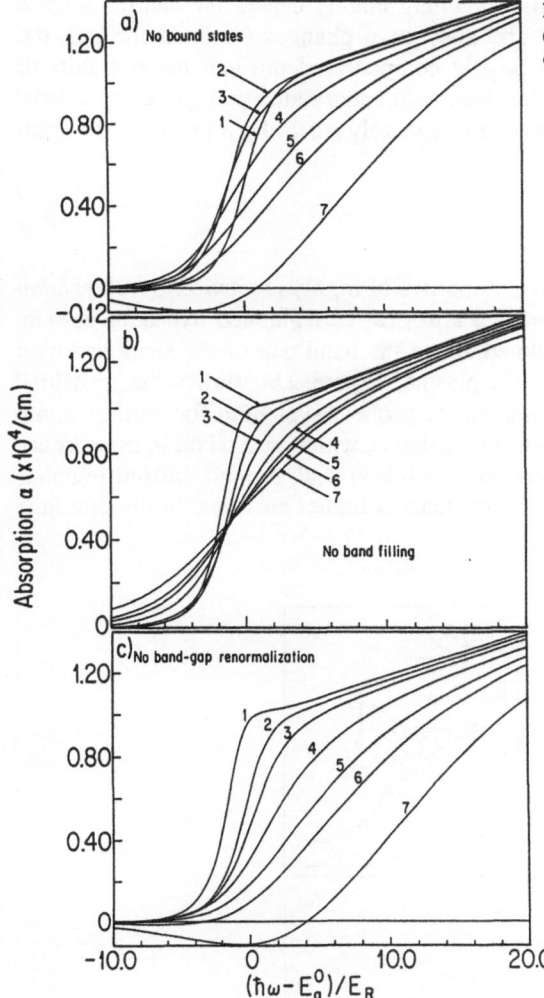

Fig. 2.8a-c. The relative contribution of different nonlinear effects from the theory

The dispersive changes are plotted in Fig. 2.7d. For frequencies below the spectral position of the excition, one has a negative Δn that drops off very slowly. A detailed analysis [2.12] shows that the relatively broad structure responsible for the tail is caused by band filling effects, whereas the superimposed sharper structure is mainly due to screening of the continuum states.

In Fig. 2.8, we have tried to distinguish between different many-body effects. The absorption spectrum is calculated by artificially turning off (a) the excitonic (bound state) contributions, (b) band filling, and (c) band-gap renormalization. The differences between Fig. 2.7c and Fig. 2.8 illustrate the relative contributions from the respective sources. Depending on the frequency regime, band filling and screening of the continuum states are the most efficient mechanisms responsible for the net dispersive changes: band filling is responsible for the broad low-frequency tails of Δn, whereas the screening mainly causes the sharp structure in the vicinity of the band gap. The absorptive changes obtained through the bleaching of excitonic states are largely compensated through the red shift of the continuum states caused by the band-gap renormalization. In other words, the negative Δn from exciton bleaching is largely offset by a positive Δn from band-gap reduction at low intensities.

2.1.2 ZnSe

In order to observe the steady-state properties of highly excited ZnSe, *Peygham-barian* et al. [2.11] used a nanosecond nitrogen-laser-pumped dye laser system. The pump beam frequency was tuned above the band gap of the semiconductor to generate directly an electron-hole plasma. Figure 2.9a shows the measured ZnSe absorption spectrum as a function of probe wavelength for various pump intensities at $T = 300$ K. The spectrum labeled 0, which was taken in the absence of the pump beam, exhibits a pronounced exciton peak around 450 nm together with the Coulomb-enhanced continuum states at higher energies. In the presence

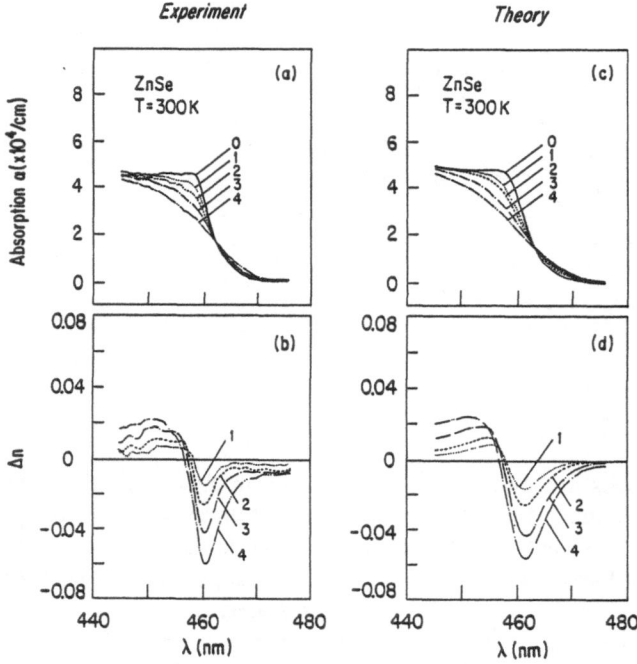

Fig. 2.9a–d. Experimental and theoretical spectra of the nonlinear absorption and nonlinear refractive index for a 0.55-μm ZnSe thin film at room temperature. (a) Experimental absorption spectra for various pump intensities: (0) no pump, (1) 16 kW/cm^2, (2) 33 kW/cm^2, (3) 76 kW/cm^2, and (4) 157 kW/cm^2. (b) Nonlinear refractive index changes, curves 1–4 are obtained from curves 1–4 in (a), respectively. (c) Calculated spectra for different electron-hole pair densities, N: (0) 10^{15} cm^{-3}, (1) 9×10^{16} cm^{-3}, (2) 1.5×10^{17} cm^{-3}, (3) 3×10^{17} cm^{-3}, and (4) 5×10^{17} cm^{-3}. (d) Calculated nonlinear refractive index changes; curves 1–4 are obtained from curves 1–4 in (c), respectively

of the pump pulse, one observes a reduction of the exciton oscillator strength and a small broadening on both sides of the exciton. At higher pump intensities, the exciton gradually disappears. The change in the index of refraction, Δn, is obtained by making a Kramers-Kronig transformation of the measured change in absorption coefficient, $\Delta\alpha$ as shown in Fig. 2.9b.

The experimental data are analyzed using the plasma theory described above with the material parameters at $T = 300\,\mathrm{K}$ given in Table 2.1. Examples of the results are shown in Fig. 2.9c. Again, the low-density absorption curve (0) is the spectrum of the unexcited crystal. The $1s$-exciton peak vanishes at higher plasma densities because of the strong screening. Since the highest carrier density in the studied situation barely exceeds the Mott density, the nonlinear mechanism like band filling and band-gap renormalization are not very important under the present conditions. In Fig. 2.10 we again show the calculated nonlinear index changes when we artificially eliminate the band filling and the collision broadening. This figure clearly indicates that exciton screening and broadening are the main nonlinear mechanisms. The close resemblance of Fig. 2.10b to 2.10a suggests that band filling does not play a major role in the nonlinear dispersion at low carrier densities. At much higher carrier densities, i.e. above the Mott density, band filling starts to become important. The calculated results obtained by applying the Kramers-Kronig transformation to the curves in Fig. 2.9c are plotted in Fig. 2.9d. For frequencies below the spectral position of the exciton, one has a negative Δn, as seen in the experiments.

Similar experiments and calculations were performed at $T = 150\,\mathrm{K}$ as shown in Fig. 2.11. The presence of the exciton resonance at low carrier concentrations is more evident due to the reduced phonon broadening. The highest pump intensities used in both the $T = 300\,\mathrm{K}$ and the $T = 150\,\mathrm{K}$ experiments correspond to pair densities close to the Mott density, such that at these highest intensities the exciton has just ionized. Again, the agreement between experiment and theory is quite good.

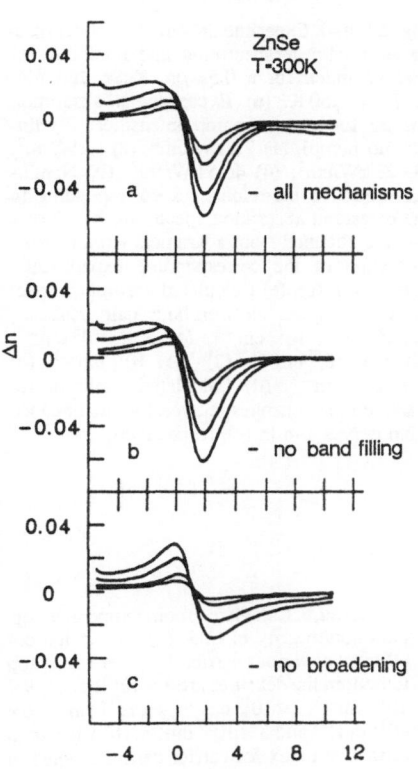

Fig. 2.10. Comparison of calculated nonlinear refractive index changes including (a) all mechanisms, (b) all mechanisms except band-filling, and (c) all mechanisms except collision broadening. The abscissa is the energy shift from the (unpumped) band-gap energy measured in units of the exciton Rydberg

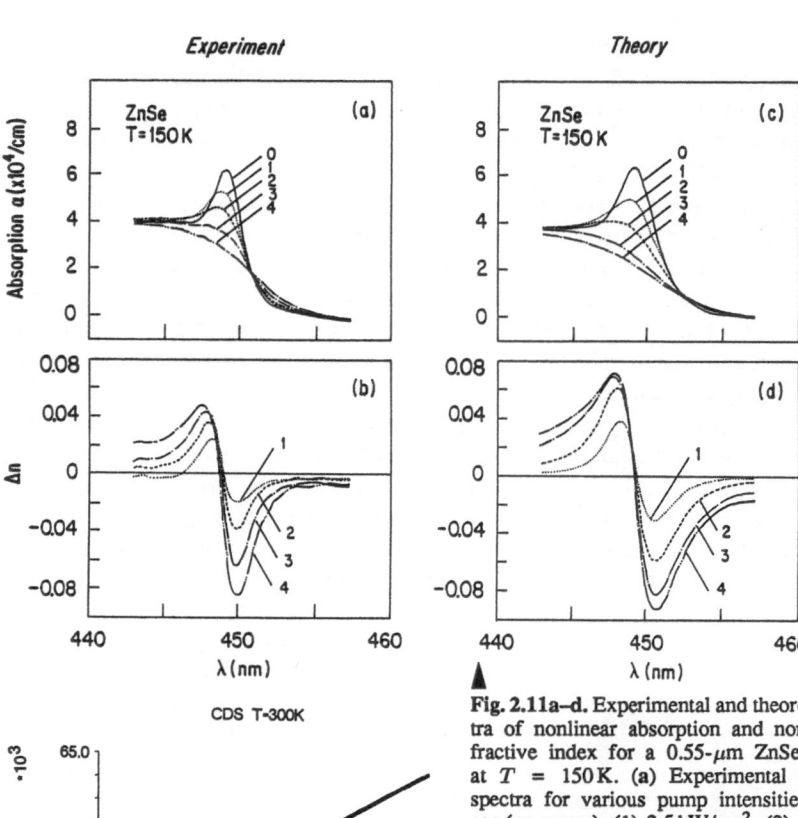

Experiment

(a) ZnSe T=150 K
Absorption α (×10⁴/cm)
0-1-2-3-4

(b) Δn

Theory

(c) ZnSe T=150 K
0 1 2 3 4

(d)

λ (nm)

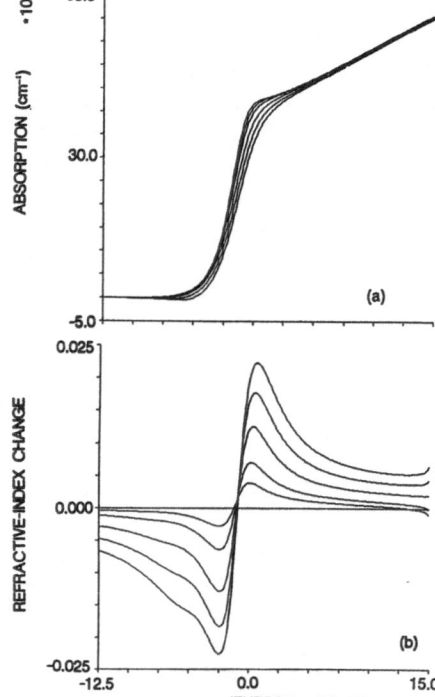

CDS T-300K

ABSORPTION (cm⁻¹) •10³

(a)

REFRACTIVE-INDEX CHANGE

(b)

(ENERGY - EG₀)/E_R

Fig. 2.11a–d. Experimental and theoretical spectra of nonlinear absorption and nonlinear refractive index for a 0.55-μm ZnSe thin film at T = 150 K. **(a)** Experimental absorption spectra for various pump intensities: (0) linear (no pump), (1) 2.5 kW/cm², (2) 9 kW/cm², (3) 23 kW/cm², (4) 46.5 kW/cm². **(b)** Nonlinear refractive index changes corresponding to the measured absorption spectra of (a). Curves 1–4 are obtained from a Kramers-Kronig transformation of the corresponding experimental data 1–4 in (a). **(c)** Calculated absorption spectra for different electron-hole pair densities, N: (0) 1×10^{15} cm⁻³, (1) 5×10^{16} cm⁻³, (2) 1×10^{17} cm⁻³, (3) 2×10^{17} cm⁻³, (4) 3×10^{17} cm⁻³. **(d)** Calculated nonlinear refractive index changes; curves 1–4 are obtained from curves 1–4 in (c), respectively

Fig. 2.12a,b. Calculated room temperature optical nonlinearity in CdS. **(a)** Absorption coefficient at various carrier densities – from top to bottom the densities are 1×10^{16} cm⁻³, 1×10^{17} cm⁻³, 2×10^{17} cm⁻³, 4×10^{17} cm⁻³, 6×10^{17} cm⁻³ and 8×10^{17} cm⁻³. **(b)** Change in refractive index for carrier densities noted in (a)

2.1.3 CdS

A comparison between theory and experiment for the case of bulk CdS has been performed by *Wegener* et al. [2.60]. Using the material parameters at $T = 300\,K$ given in Table 2.1, the absorption and refractive index spectra have been computed and the results are shown in Fig. 2.12. Similar to ZnSe, in room temperature CdS, exciton screening at low carrier densities and band filling for elevated densities have been identified as the dominant sources of the electronic nonlinearity. At lower temperatures ($T = 30$–$60\,K$), it has been shown that the band-gap reduction becomes quite prominent, as was shown in Fig. 2.2. As discussed earlier, for frequencies somewhat below the exciton resonance, the shrinking band gap causes the absorption to increase with increasing carrier density.

In addition to free excitons, the bound excitons may also provide optical nonlinearity. Bound exciton resonances are usually present in semiconductors as a result of the attachment of free excitons to impurities (either donors or acceptors). A bound exciton is energetically located below the free exciton by the binding energy of the exciton to the impurity. Bound excitons are stationary at the impurity site and cannot move through the crystal, in contrast to free excitons. The I_2 exciton is bound to an impurity donor in CdS. An experiment that was performed at $2\,K$ with cw blue light showed [2.61] that the narrow bound exciton may be saturated with very little power. The large oscillator strength of the bound exciton may lead to a radiative lifetime of $\cong 0.5\,ns$.

2.1.4 CuCl

CuCl is a wide band-gap semiconductor with a band-gap energy of $\cong 3.4\,eV$ at $4\,K$. The free exciton in CuCl is highly bound with a binding energy of $\cong 190\,meV$ and a small Bohr radius of $\cong 7\,\text{Å}$. Exciton screening and electron-hole plasma behavior under high excitation intensities have been investigated in this material using subpicosecond pulses [2.62, 63]. The lowest biexciton level is a totally symmetric state of spin-paired electrons and holes with a total energy at $k = 0$ of $6.3722\,eV$ or $3.1861\,eV$ per electron-hole pair. The biexciton binding energy is about $30\,meV$. The first observation of biexcitons was made in CuCl through an analysis of the luminescence spectrum using band-to-band excitation [2.64]. Biexcitons were then verified through direct two-photon absorption [2.65–70], degenerate four-wave mixing [2.71, 72], phase conjugation [2.73], and polarization rotation measurements [2.74]. Figure 2.13 displays the absorption spectra of CuCl grown on NaCl [2.75, 76]. The biexciton two-photon absorption at $3890\,\text{Å}$ is located on the low energy tail of the exciton absorption line. In a simple approximation, the optical spectrum around the biexciton resonance may be calculated from the dielectric function [2.77, 78]

$$\varepsilon(\omega) = \varepsilon_\infty + \frac{(\varepsilon_0 - \varepsilon_\infty)E_x^2}{E_x^2 - (E - i\gamma_x)^2 - \Omega^2(E)} \quad \text{where} \tag{2.5}$$

$$\Omega^2(E) = \frac{4n_p|M|^2 E_x(E_{xx} - E - i\gamma_{xx})}{(E_{xx} - i\gamma_{xx})(E - xx - 2E - i\gamma_{xx})} . \tag{2.6}$$

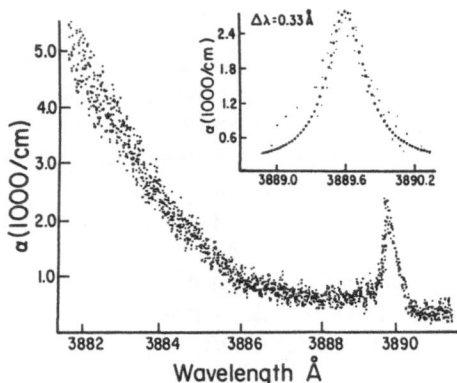

Fig. 2.13. Absorption spectra for 4.0-μm thin films of CuCl grown on NaCl

Here, ε_0 and ε_∞ are the dielectric constants at low and high frequencies, which are related by the transverse–longitudinal splitting δ

$$\delta = E_x \left[\sqrt{\varepsilon_0/\varepsilon_\infty} - 1 \right] \tag{2.7}$$

where E_x and E_{xx} and γ_x and γ_{xx} are the exciton and biexciton energies and their full widths at half maximum, respectively, n_p is the polariton density, and M is the matrix element for biexciton formation. By analyzing the above equations, the absorption coefficient and Δn were obtained as a function of energy for different pump intensities [2.47, 78–83].

Biexcitonic optical nonlinearities can be studied by different methods. Here, we discuss the measurements of the nonlinear index of refraction using a polarization rotation technique [2.74, 84]. Selection rules for direct biexciton two-photon absorption allow production of a biexciton state with two circularly polarized light beams with opposite polarization directions, i.e. left- and right-circularly polarized beams can produce a biexciton state, while the transition for two left- or two right-circularly polarized beams is forbidden. Therefore, if two beams, a pump beam with left-circular polarization and a linearly polarized probe beam, each tuned near half the biexciton energy, are focused on the sample, the right- and left-circularly polarized components of the probe suffer different amounts of phase shift and absorption. Thus, the linear polarization of the probe beam changes to an elliptical one. The angle of rotation of the polarization axis, $\theta = \Delta n \omega d/2c$ where ω is the frequency of the light and d is the sample thickness. Such experiments yielded [2.74] a variation of Δn as a function of frequency near the biexciton resonance. The maximum value for Δn was about 0.002 with 1 MW/cm^2 pump intensity.

2.1.5 InSb, InAs, and CdHgTe

In narrow gap materials such as InSb, InAs, and CdHgTe, band filling is the dominant source of the optical nonlinearity [2.85–87]. The band gap of these materials ranges from 10 to 3 μm (see Table 2.1 for other material parameters), and the exciton resonances are usually not resolved. This, however, does not imply that these materials exhibit a free-carrier nonlinearity. As is clearly visible in our Fig. 2.3 for InSb, the absorption spectrum does not exhibit a $\sqrt{E}$-type behavior, but rather shows a steep increase around the band edge. This steep increase is due to the exciton that is not spectrally resolved and the electron-hole Coulomb enhancement effects discussed above. In other words, these narrow-gap materials behave quite similarly to the semiconductors with a wider gap, the major difference being the small exciton binding energy inhibiting the spectral resolution of the exciton.

2.1.6 Thermal Nonlinearities in GaAs, CdS, ZnSe, and ZnS

The index of refraction of a material can be changed by varying the temperature. A laser beam can be absorbed and can locally heat the material, causing a thermal index change. A resonance line can also be shifted by changes in temperature, resulting in a nonlinear index effect. This kind of nonlinearity has been employed in various materials such as ZnS, ZnSe [2.88, 89], GaAs [2.26, 90] and CdS [2.91] for obtaining optical bistability.

2.2 Optical Nonlinearities in Low-Dimensional Semiconductors

2.2.1 Quantum Wells

Quantum confinement effects in optically excited semiconductor microstructures arise if at least one spatial dimension of the material becomes comparable to or smaller than the exciton Bohr radius, which is the characteristic length scale of an electron-hole pair. Well-known examples of such semiconductor systems are the multiple-quantum-well structures made of alternating layers of active and transparent materials. Laser excitation in the appropriate frequency regime generates electron-hole pairs within the quasi-two-dimensional active layers. These layers provide confinement in one space dimension which is already sufficient to greatly enhance excitonic effects. In two dimensions, the binding energy of the excitons with principal quantum number n is given by [2.92] $E_n = E_R / n + \left(\frac{1}{2}\right)^2$, $n = 0, 1, 2, \ldots$, where E_R is the exciton Rydberg, whereas, in three dimensions, $E_n = E_R / n^2$, $n = 1, 2, \ldots$. The binding energy of the lowest exciton state is therefore four times larger in two than in three dimensions. This particular feature makes exciton effects easily observable even at room temperature.

In this section, we will concentrate on the nonlinearities of two types of quantum well systems: GaAs/AlGaAs and ZnSe/ZnMnSe.

(a) GaAs-AlGaAs Multiple Quantum Wells. Modern crystal growth techniques such as Molecular Beam Epitaxy (MBE) and Metal-Organic Chemical Vapor

Deposition (MOCVD) make it possible to manufacture high quality Multiple-Quantum-Well (MQW) structures with precise layer thicknesses and high uniformity [2.93]. The excitonic effects can be seen easily at room temperature in MQWs as a result of the confinement effect (which results in a smaller exciton Bohr radius and consequently a larger exciton binding energy, leading to an enhanced exciton oscillator strength) and small LO phonon broadening. (The coupling to LO phonons does not significantly change from bulk GaAs to MQWs [2.94].)

Some of the nonlinear effects in quasi-two-dimensional (2d) MQWs and their comparison with three-dimensional (3d) bulk materials may be summarized as follows (for more detailed discussions see, e.g. [2.8, 41, 52, 95]). Coulomb screening is weaker in quasi-2d structures than in 3d. Intuitively, one can understand this fact by noting that screening is a spatial rearrangement of the electrons and holes. The possibilities for such a rearrangement are very restricted in quantum wells since the carriers can only move within the quasi-two-dimensional layers, hardly influencing those "field lines" which are within the barrier material. However, one knows from the many-body analysis of the band-edge nonlinearities in semiconductors that screening is not the only effect which leads to an effective weakening of the electron-hole Coulomb attraction. If one derives the effective Wannier equation for the exciton using many-body theory [2.46, 51, 96], one finds

$$
\begin{aligned}
i\frac{\partial}{\partial t}\phi(x) = & - H_{kin}(x)\phi(x) + \int dr V_s(r)\phi(r)[\delta(x-r) \\
& - n_e(x-r) - n_h(r-x)]
\end{aligned}
\tag{2.8}
$$

where $\phi(x)$ is the radial wavefunction for the relative motion of an electron-hole pair with electron-hole separation x, H_{kin} is the Hartree Hamiltonian of the electron and hole and the last term describes the effective Coulomb attraction between electron and hole, where $n_{e/h}(x)$ gives the probability of having an electron/hole at space point x. The details of this equation are not relevant for our discussion here. It is only important to note that the factor $[\delta(x-r) - n_e(x-r) - n_h(r-x)]$ multiplies the screened Coulomb potential $V_s(r)$. This *phase-space filling factor* arises because an exciton is made out of an electron and a hole, each of which obeys the Pauli exclusion principle. Hence one can use only those electron-hole states for the exciton that are not yet occupied by free electrons and holes. Thus the presence of free electrons and holes reduces the effective electron-hole attraction not only through the screening (in V_s) but also through phase-space filling. Both effects are always present simultaneously but their relative importance changes with the dimension of the system. In three dimensions at not too low temperatures, one can obtain good agreement between theory and experiment if one neglects the phase-space filling because the plasma screening reduces the electron-hole attraction so rapidly that the additional reduction due to phase-space filling is a relatively minor effect. In quasi-two-dimensional systems, however, phase-space filling yields an important contribution to the nonlinearity, mainly because the efficiency of screening is reduced and the phase space is more readily filled.

Even though the physical processes causing renormalization of the optical spectra near the band edge of bulk and MQW systems are somewhat different, the overall features of the nonlinearities themselves are quite similar. The magnitude of the nonlinear refractive index has been measured for MQWs using degenerate four-wave mixing (see for example [2.97, 98]), pump-probe spectroscopy [2.99, 100] and nonlinear interferometry [2.59]. The nonlinear absorption and refraction have been measured as a function of well size at room temperature [2.58, 100]. In these experiments, a nitrogen-pumped dye laser operating at 816 nm was used for the pump beam. The probe consisted of the spectrally broadband luminescence emitted by a cell containing infrared dyes. The nitrogen laser synchronously pumped both the dye laser and the dye cell, which subsequently delivered pulses of 3-ns full width at half maximum (FWHM). The spot diameter of the pump beam at the sample was $\cong 200 \ \mu$m, allowing neglect of the diffusive decay time in comparison to the recombination time. The probe beam diameter was $\cong 50 \ \mu$m.

Four different MBE-grown samples were used for this research: 76, 152 and 299 Å GaAs/AlGaAs MQWs and bulk GaAs. The first MQW sample consisted of 63 layers of 76-Å-thick GaAs. Each GaAs layer is followed by an 81-Å-thick layer of $Al_{0.37}Ga_{0.63}As$. The total GaAs thickness of this sample was 0.48 μm. The next sample consisted of 100 layers of 152-Å-thick GaAs and 104-Å-thick $Al_{0.33}Ga_{0.67}As$ barriers. The third sample had 61 layers of 299-Å-thick GaAs and 98-Å-thick $Al_{0.36}Ga_{0.64}As$, resulting in a total GaAs thickness of 1.8 μm. The bulk GaAs consisted of a 2.05-μm GaAs layer sandwiched between two AlGaAs layers. These four samples are called 76, 152, 299 Å, and bulk, respectively, throughout this discussion.

The absorption changes for the various pump intensities were measured in the vicinity of the band edge at room temperature. The pump wavelength was fixed above the band edge at 816 nm. The same pump beam intensities were used for all four samples. Figure 2.14a shows the nonlinear absorption spectra for bulk GaAs. The exciton peak, which is barely resolvable in bulk, is more prominent in the 299 Å sample, as shown in Fig. 2.14b. For low pump intensity, exciton saturation dominates. For high intensity, the nonlinear behavior in the 299 Å sample is similar to that of bulk GaAs because of the relatively weak confinement. In Fig. 2.14c and d, similar curves are displayed for the 152 and 76 Å samples. The increase in the excitonic absorption is now very apparent. Notice that separation of the heavy- and light-hole excitons is very clear in the 76 Å sample in contrast to the 152 Å sample. The corresponding variation in the refractive index is shown in Fig. 2.15.

Curves of Δn versus intensity for all four samples are plotted in Fig. 2.16a. Note from Fig. 2.16a that, for a given Δn, a lower pump intensity is required for the 76 Å MQW sample than for the bulk sample. These results indicate that the smaller MQWs have larger nonlinear refractive indices. However, we note that the comparison of intensity-dependent index changes is somewhat misleading.

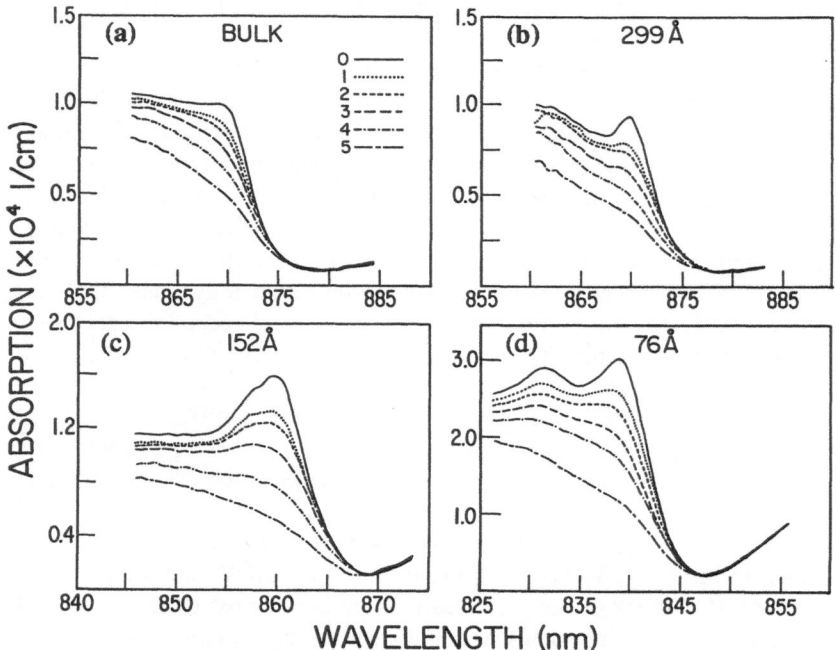

Fig. 2.14. Experimental room-temperature absorption spectra for: **(a)** bulk GaAs; **(b)** 299 Å MQW; **(c)** 152 Å MQW; **(d)** 76 Å MQW. The curves labeled in the figure represent pump beam intensities of (0) 0; (1) 670; (2) 1270; (3) 2650; (4) 5400; (5) 11 700 W/cm^2

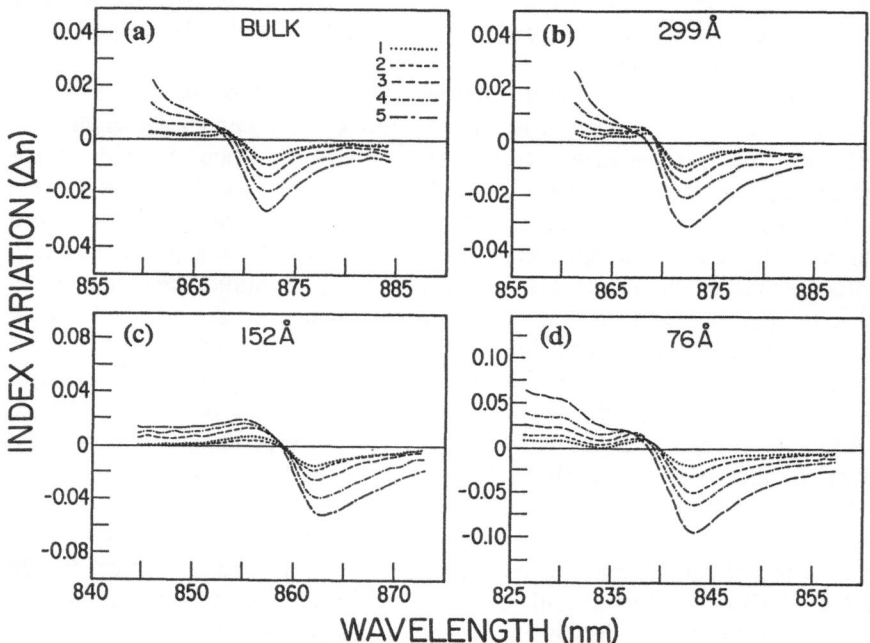

Fig. 2.15. Refractive index changes corresponding directly to the measured changes in the absorption spectra for **(a)** bulk GaAs; **(b)** 299 Å MQW; **(c)** 152 Å MQW; **(d)** 76 Å MQW

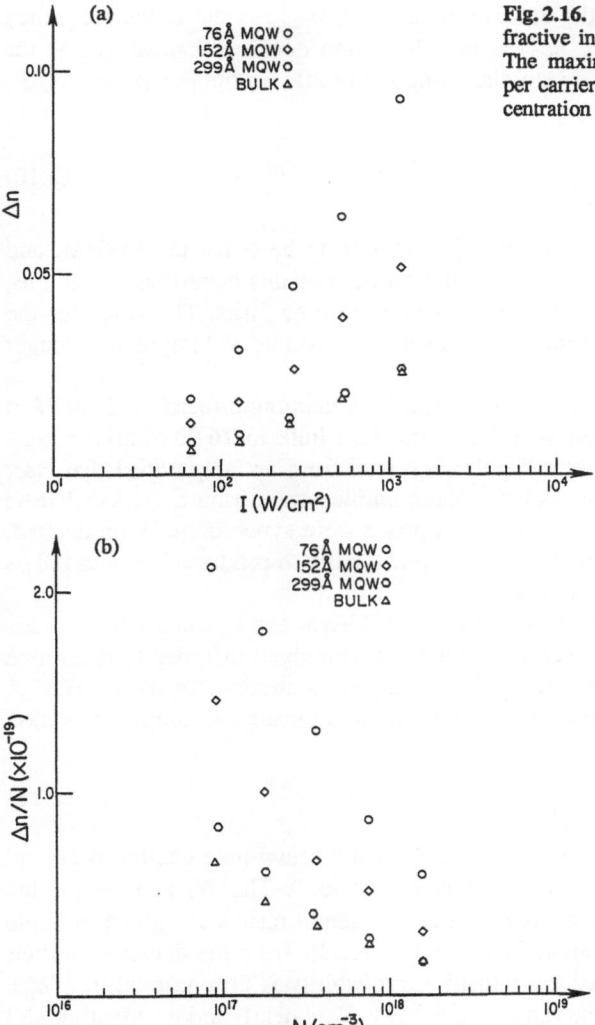

Fig. 2.16. (a) The maximum change in refractive index vs pump beam intensity. (b) The maximum change in refractive index per carrier concentration vs the carrier concentration

For resonant excitation, changes of the optical material properties depend on the excitation intensity only through the carrier density. Moreover, the various samples have a different absorption at the pump frequency so that the same excitation generates a different number of carriers. Therefore, to obtain a more fundamental comparison of the material nonlinearities themselves, the index change per excited carrier, $\Delta n/N$, is compared for these samples.

Figure 2.16b shows the maximum $\Delta n/N$ for a given carrier concentration. The carrier concentration is obtained using the simple rate equation

$$\frac{dN}{dt} = \frac{\alpha(\omega_{\text{pump}})I}{\hbar\omega_{\text{pump}}} - \frac{N}{\tau} \tag{2.9}$$

where τ is the lifetime of the electron-hole pairs, ω_{pump} is the radial frequency of the pump beam, and $\alpha(\omega_{pump})$ is the absorption coefficient measured at the pump frequency. Assuming a Gaussian temporal profile for the pump beam, (2.9) yields

$$N(t) = \frac{e^{t/\tau}\alpha(\omega_{pump})I_0}{\hbar\omega_{pump}} \int_{-x}^{t} dt' e^{-t/\tau} e^{-[(1.6651/\text{FWHM})t]^2} \tag{2.10}$$

where I_0 is the peak pump intensity, x is chosen to be twice the FWHM, and t is the delay time between the pump and probe. For this experiment, t is 1 ns, FWHM is 3 ns, and the carrier lifetime is chosen to be 20 ns. The value for the carrier lifetime does not significantly influence our results as long as it is longer than the 3 ns pulse width.

As shown in Fig. 2.16b, $\Delta n/N$ increases by a maximum factor of 3 for $N \cong 10^{17}$ cm^{-3} as the MQW well size decreases from bulk to 76 Å. Similar results were obtained using microsecond pulses derived from an Ar-pumped dye laser operating as the pump beam and the photolominescence from a GaAs/AlGaAs MQW as the probe beam. The pump and probe were synchronously modulated, at a repetition rate of 5 kHz, by acousto-optical cells to produce 1 μs and 0.8 μs rectangular pulse trains, respectively.

As can be seen from Fig. 2.14, the excitons bleach and broaden with increasing carrier concentration. Several models were employed in order to determine the excitonic absorption saturation with carrier concentration for the MQWs. A reasonable fit to the experimental data was obtained using the simple saturation model

$$\alpha(N) = \frac{\alpha_0}{1 + N/N_s}, \tag{2.11}$$

where α_0 is the linear absorption coefficient at the heavy-hole exciton peak and N_s is an empirical "saturation carrier concentration". The N_s and α_0 for the MQW samples obtained from fits of the experimental results are given in Table 2.2. The larger α_0 for the smaller-well MQWs results from the decrease in their exciton Bohr radii, a_B (since the transition probability is proportional to $1/a_B^3$). These results indicate that the saturation density N_s is nearly independent of well size and that the factor of three increase in the optical nonlinearities is mainly due to the factor of three increase in the exciton absorption.

(b) Semiconductor Doping Superlattices (n-i-p-i Structures). Semiconductor doping superlattices or n-i-p-i crystals are periodic structures of n-type intrin-

Table 2.2. Values of α_0 and N_s determined by least squares fits

MQW Sample	$\alpha_0 [\times 10^4$ cm$^{-1}]$	$N_s [\times 10^{17}$ cm$^{-3}]$
76 Å	2.7	9.0
152 Å	1.4	7.2
299 Å	0.8	11.1

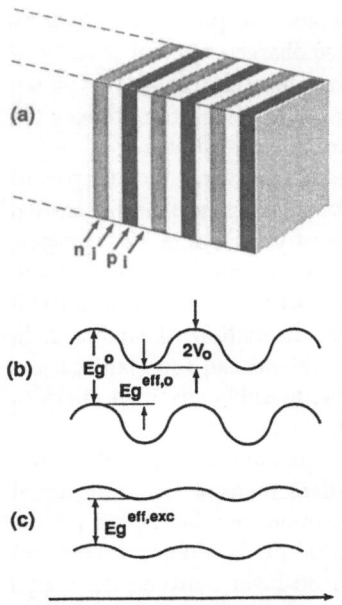

(a)

(b)

(c)

Direction along periodicity

Fig. 2.17. (a) Schematic structure of a n-i-p-i superlattice. (b) Band diagrams for no additional injected carriers, (c) Band diagram with additional injected carriers

sic (undoped), p-type intrinsic materials [2.101–103]. Figure 2.17a shows the schematic for such a structure in real space. A donor may transfer an electron to an acceptor, making both types of impurities charged. The ionized impurities cause a buildup of a periodic space charge potential in the doping layers which modulates the conduction and valence band edges. The modulation of the electronic bands has the strength of $2V_0$. The space charge potential is parabolic, as shown in Fig. 2.17b, in contrast to the rectangular potentials in compositional superlattices that were described in the previous section. In compositional superlattices such as GaAs/AlGaAs, the superlattice potential originates from different band gaps of the components. The energy gap of the n-i-p-i structure, E_g^{eff}, is the energy separation between the top of the valence band in the p-layer and the bottom of the conduction band in the n-layer [2.103, 104]

$$E_g^{\text{eff}} = E_g^0 - 2V_0 , \tag{2.12}$$

where E_g^0 is the band gap of the host material. The built-in potential (electric field) makes the electrons go to the bottom of the conduction band in the n-layer and the holes accumulate at the top of the valence band p-layer. Therefore, the electrons and holes are spatially separated, which results in longer recombination lifetimes.

If additional carriers are injected in the material, either by light absorption or through application of electrical currents, the impurity space charge potential is partially compensated since electrons and holes partially neutralize the positively charged donors and negatively charged acceptors, respectively. Thus, by increasing the excess carrier concentration, the space charge potential is reduced and the effective n-i-p-i band gap increases, approaching the band gap of the host material E_g^0 as schematically displayed in Fig. 2.17c. This effect makes the doping superlattices tunable in band gap and recombination lifetime.

The description given above represents a simple picture where the energies of the electronic states (valence-band and conduction-band states) are merely shifted by the built-in space charge field. The confinement of the carriers in the super-lattice direction was ignored. However, for thin layers, the quantum confinement energy becomes significant and the motion of particles in the direction normal to the layers becomes quantized. This leads to the formation of subbands in the conduction and valence bands. In contrast to compositional superlattices, the parabolic potentials in doping superlattices give rise to subbands with energies and spacings that follow harmonic oscillator levels.

So far we have only considered simple n-i-p-i structures in which the modulation is due to dopants. In hetero-n-i-p-i superlattices there is a compositional variation in addition to doping modulation. For example, one hetero-n-i-p-i superlattice uses a larger-gap material (such as AlGaAs) as the doping layers and a smaller-gap material (such as GaAs) as the intrinsic layer between the doped layers. In these structures, there is a potential associated with the spatial variation of the band structure of the two different materials in addition to the space charge potential. The hetero-n-i-p-i structures combine the unique features of the compositional and doping modulated superlattices.

N-i-p-i superlattices have been grown from various materials such as GaAs [2.103], InP [2.105], InGaAs [2.106], Si [2.107], and PbTe [2.108]. Here, we only concentrate on optical nonlinearities of GaAs-based structures. Recently, interesting optical nonlinearities have been observed in a GaAs/AlGaAs hetero-n-i-p-i superlattice [2.109]. The reported structure that was grown by MOCVD used a MQW between doped layers of $Al_{0.32}Ga_{0.68}As$ with dopant concentrations of $p \cong n \cong 2 \times 10^{18}$ cm^{-3}. Optical absorption in the quantum wells creates electrons and holes which drift to the n- and p-type regions, respectively. These excess carriers compensate the built-in field of charged impurities which modify the absorption in the quantum wells. Absorption changes of more than 2000 cm^{-1} were observed with low intensities of 375 mW/cm^2. This corresponded to an index change of $\cong 0.02$ or a value for $\Delta n / I$ of 50 cm^2/kW, which is more than two orders of magnitude larger than that in a GaAs/AlGaAs MQW. Even though the response time of this nonlinearity was not reported, it is most likely that the larger nonlinear index had been obtained at the cost of a slower speed. The spatial separation of carriers usually increases the carrier lifetime to microseconds and even slower. However, the slower speed may be acceptable for some applications such as parallel processing; certainly the lower operating powers are most welcome for parallel operation.

(c) ZnSe/ZnMnSe Superlattices. The ZnSe/ZnMnSe structures are among the wide-gap II-VI superlattices. They have the potential to lead to blue light emitters in the future [2.110]. Optical nonlinearities of these strained superlattices have been measured recently [2.110, 111]. Two-nanosecond pulses were used to measure the saturation intensity of a superlattice consisting of 67 periods of 73 Å ZnSe wells and 185 Å $Zn_{0.49}Mn_{0.51}Se$ barriers. A saturation intensity of $\cong 1.3\,kW/cm^2$ at 77 K was reported.

2.2.2 Quantum Wires, Quantum Dots, and Semiconductor Doped Glasses

(a) **Quantum Wires.** Stimulated by the interest in quantum-well structures, researchers have been trying to further reduce the dimensionality of the semiconductor systems. It has become customary to use the term *quantum wires* for structures where the electron-hole pairs are confined in two space dimensions. Attempts fo fabricate quantum wires have been made by etching small stripes out of quantum well material [2.34, 36, 112] or by using indirect methods to effectively confine the carriers to quasi one dimension [2.133]. For the sake of completeness, we wish to mention that optical nonlinearities in effectively one-dimensional systems can also be studied in some organic materials, such as polydiacetylene [2.114]. A discussion of organic semiconductors, however, goes beyond the scope of this article. At present, we are not aware of published experimental results reporting optical nonlinearities in fabricated quantum wires. This is basically due to the fabrication problems associated with the small sizes and the required size uniformity. However, other interesting effects have been seen in quantum wires and some theoretical investigations have been performed. For example, a strong increase of the ionization energy of shallow impurities in quantum-well wires has been predicted [2.115] and interesting modifications of the excitonic and biexcitonic effects in quasi one dimension have been suggested [2.39]. The following is a summary of some of the observations and calculations.

Clear evidence of two-dimensional confinement in (AlGa)As/GaAs quantum well wires, with sizes of a few nanometers, has been observed [2.116, 117]. The samples were grown by MBE on tilted substrates. The evidence for quantum confinement was obtained in the strong anisotropic photoluminescence and excitation spectra. The ratio of the electron–light-hole exciton peak intensity (I_{e-lh}) to the electron–heavy-hole exciton peak intensity (I_{e-hh}) was found to be dependent on the light polarization. For light polarized with its electric field vector nearly parallel to the wires, the heavy-hole peak, I_{e-hh}, was more intense than the light-hole peak, I_{e-lh}, with $I_{e-lh}/I_{e-hh} \cong 0.4$; whereas, when the incident light was polarized nearly perpendicular to the wires, the light-hole peak was more intense than $I_{e-lh}/I_{e-hh} \cong 1.8$. This anisotropy agreed rather well with theory and indicated two-dimensional quantum confinement.

GaAs/GaAlAs quantum wires fabricated by ion-implantation-induced inter-diffusion have also been studied [2.118]. The photoluminescence spectroscopy from these samples showed that the emission from the quantum wires (masked region) was shifted relative to the unmasked region of the sample. Furthermore, stimulated emissions have been observed from two-dimensionally confined carriers in quantum wires [2.119]. The structures were fabricated by the lateral patterning technique [2.120] where GaAs/AlGaAs single quantum well laser heterostructures were grown on V-grooved GaAs substrates using organometallic chemical vapor deposition. The fabricated structure was 100 Å thick and less than 1000 Å wide GaAs. Transitions between quasi-one-dimensional subbands were measured at room temperature. Above threshold, lasing occurred at one of these subband transitions.

On the theoretical front, *Banyai* et al. [2.39] studied the quasi-one-dimensional confined system by assuming that the wavefunctions for the electrons and holes can be written as a product wavefunction between the ground state in the two "frozen" dimensions and a function of the remaining single degree of freedom. Thus, the quantum mechanical problem reduces to the determination of the wavefunction in one dimension with an effective Coulomb interaction, which is just the normal Coulomb interaction averaged over the two "frozen" degrees of freedom. This procedure is not only very reasonable from a physics point of view, but it is also necessary, since the true Coulomb problem in one dimension is pathological [2.121], i.e. the electron-hole binding energy diverges, and the higher levels become degenerate as one tries to extrapolate to zero radius. These problems show that the purely one-dimensional treatment is insufficient, and one is compelled to retain the explicit radius dependence of the effective interaction and therefore of the quasi-one-dimensional binding energies and wavefunctions. In fact, since the confining potential is of finite height, as the well width is reduced, the ground state is pushed closer to the continuum. Thus if the wire radius is reduced too much, it becomes more and more probable that the electron and/or hole tunnel out of the wire, and eventually the bulk properties must be recovered. Such an analysis has been made by *Bryant* [2.115] for impurities centered on the wire axis. According to his estimates in GaAs/GaAlAs, this phenomenon becomes important for diameters less than the exciton radius. *Banyai* et al. [2.39] find that the excitonic binding energies of quantum-well wires are about five times the bulk value, and the biexcitonic binding energies in a quasi-one-dimensional quantum-well wires are broadly similar to those found in quantum wells.

(b) Quantum Dots. We denote as quantum dots all those semiconductor microstructures which confine the laser-excited electron-hole pairs in all three space dimensions. The characteristic length (radius R) of quantum dots is comparable to the exciton Bohr radius $a_B = \hbar^2 \varepsilon_2 / \mu e^2$, where $1/\mu = 1/m_e + 1/m_h$ and ε_2 is the background dielectric constant of the semiconductor material. For microcrystals with radius R in the range $l_0 \ll R \cong a_B$, where l_0 is the lattice constant of the semiconductor, the band-structure properties are determined by the periodic lattice. That is, one may assume that the effective mass approximation holds, with the electron and hole having effective masses m_e and m_h, respectively. However, optically excited electron-hole pairs with the characteristic length a_B are influenced by the small size of the microcrystals, leading to quantum confinement effects.

The theoretical investigations of quantum confinement in semiconductor microcrystallites have been pioneered by *Efros* and *Efros* [2.37] and by *Brus*

[2.38, 122]. The Hamiltonian of the system for one electron-hole pair can be written as

$$H^1 = -\frac{\hbar^2}{2m_e}\nabla_e^2 - \frac{\hbar^2}{2m_h}\nabla_h^2 + V_C \,, \tag{2.13}$$

where the first two terms are the kinetic energies of the electron and the hole, respectively, and V_C is the Coulomb interaction potential. The Schrödinger equation given by this Hamiltonian with spherical boundary conditions may be analytically solved in the absence of the Coulomb interaction. The resulting energy eigenvalues and wavefunctions are then given by

$$E^1(N_e, N_h) = E_g + \frac{\hbar^2}{2m_e}\left(\frac{\propto_{n_e l_e}}{R}\right) + \frac{\hbar^2}{2m_h}\left(\frac{\propto_{n_h l_h}}{R}\right)^2 \,, \tag{2.14}$$

$$\phi^1(N_e, N_h) = \zeta(n_e, l_e, m_e, r_e)\zeta(n_h, l_h, m_h, r_h) \,, \tag{2.15}$$

where $N \equiv \{n, l, m\}$ and

$$\zeta(n, l, m, r) = \sqrt{\frac{1}{4\pi R^3}} j_l \left[\frac{\propto_{nl}(r/R)}{j_{l+1}(\propto_{nl})}\right] Y_l^m(\theta, \phi) \,.$$

Here, j_l is the lth-order spherical Bessel function, the $Y_l^m(\theta, \phi)$ are the spherical harmonics, and $\propto_{nl}$ is the nth root of the lth-order Bessel function ($\propto_{1s} = 3.14$, $\propto_{1p} = 4.49$, $\propto_{1d} = 5.76$, etc.).

Boundary conditions $\zeta(r = 0) = $ finite and $\zeta(r = R) = 0$ are adopted in the above solutions. Equation (2.14) shows that the absorption "edge" shifts to higher energies as the crystal size, R, decreases. Furthermore, it states that the energy spectrum consists of a series of lines corresponding to the electron-hole transitions. Figure 2.18 exhibits the schematic representation of one-electron-hole-pair states where $1s$, $1p$, $1d$, etc. refer to various values of quantum numbers n, l, and m in (2.14 and 15). (For example, $1s$ corresponds to $n = 1$, and $l = 0$.) The selection rules for the dipole-allowed interband transitions are $\Delta l = 0$ in

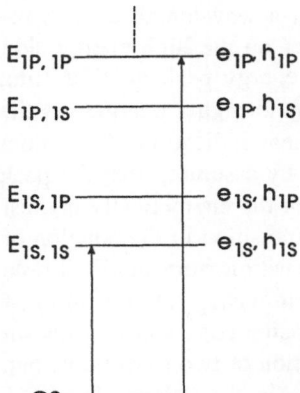

$E_{1P, 1P}$ —— $e_{1P} \cdot h_{1P}$

$E_{1P, 1S}$ —— $e_{1P} \cdot h_{1S}$

$E_{1S, 1P}$ —— $e_{1S} \cdot h_{1P}$

$E_{1S, 1S}$ —— $e_{1S} \cdot h_{1S}$

GS

Fig. 2.18. Schematic representation of the one-pair transitions in a semiconductor quantum dot. The notation e_{1s}, h_{1p} etc. refers to the electron and the hole being in $1s$ and $1p$ states, respectively

the absence of Coulomb interaction. For example, the E_{1s-1s} transition where electron and hole are both of $1s$-type is allowed.

When the Coulomb interaction is included, the problem can no longer be solved analytically and a numerical approach is needed. The absolute value of the one- and two-pair states is only weakly shifted by Coulomb effects, since the kinetic energy terms dominate for dots with $R \cong a_0$. The one- and two-pair-state wavefunctions are, however, modified enough to strongly influence the nonlinear effects. The selection rules stated earlier are no longer valid, and transitions with $\Delta l \neq 0$ are then allowed [2.43].

There has been a growing interest in searching for systems that exhibit such three-dimensional confinement effects and in understanding their behavior. A number of laboratories have attempted to fabricate quasi-zero-dimensional structures using various techniques, including colloidal suspension of semiconductor particles [2.38, 122], electron-beam lithography [2.34–36], and semiconductor microcrystallites in glass matrices [2.29, 31, 32, 123–127].

It has been shown that special glasses doped with CdS, CdSe, CuCl, or CuBr crystallites [2.29] can be fabricated that clearly exhibit quantum confinement. The microcrystallites in these glasses form out of the supersaturated solid solution of the basic constituents originally brought into the glass melt. The crystallites are more or less randomly distributed in the glass matrix. *Ekimov* et al. [2.29] report crystallite growth following the Lifshitz-Slyozov growth law [2.128] $R \propto t^{1/3}$, where R is the crystallite size and t is the duration of the heat treatment during which the crystallites actually grow. Average crystallite sizes from around $10 \, \text{Å}$ up to several $100 \, \text{Å}$ have been obtained.

In this chapter, we summarize some of the results of our femtosecond experiments performed on CdSe microcrystallites in borosilicate glass, heat treated at a temperature of 600°, 650°, or 700°C for 30 mins [2.129, 130]. Based on small-angle X-ray scattering data, the average crystallite radii of these samples were determined to be 26, 38, and 61 Å, respectively.

The linear absorption spectrum of the 600°C sample at 100 K is shown by the dotted line in Fig. 2.19. The series of discrete quantum-confined electron-hole states leads to absorption spectra which consist of a series of resonances. The linear absorption spectrum in Fig. 2.19 has a peak at a wavelength of approximately 555 nm, labeled A, a distinguishable shoulder on the high energy side of that peak at $\cong 527$ nm, labeled B, and a higher energy peak at $\cong 462$ nm, labeled C^*. As reported before, these transitions shift to higher energies as the crystallite sizes are reduced [2.31, 37]. They are a clear indication of quantum confinement. The data can be consistently explained by assuming that the peak A at 555 nm originates from the $1s$ transition between the energetically highest heavy-hole state to the lowest electron state ($E_{1s_A,1s}$ transition in the notation of Fig. 2.18) and the shoulder B originates from the transition from the light-hole state in the B valence band to the lowest electron state ($E_{1s_B,1s}$). The inset of Fig. 2.19 schematically shows these transitions. The rising edge and its peak on the high-energy side, C^*, is most probably a combination of two transitions: one being the transition from the spin-orbit split-off state in the valence band, C,

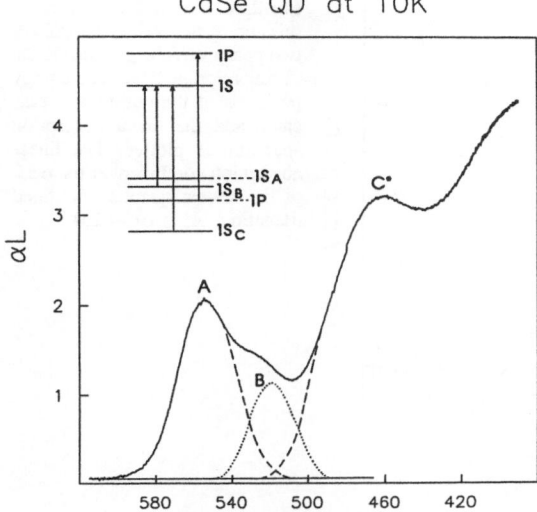

CdSe QD at 10K

Fig. 2.19. Linear absorption spectrum of the 600°C sample. The dashed lines are the decomposition of the A, B, and C^* transitions. The inset shows the schematic of the valence and conduction subbands

to the lowest electron state $E_{1s_C,1s}$, and the other being the transition between higher quantum confined states $E_{1p,1p}$.

The dashed lines in Fig. 2.19 show the decomposition of the A and B peaks obtained from the linear absorption spectrum, assuming the indicated tails of the C^* and A transitions. The ratio h_B/h_A of the peak absorption coefficients for B and A transitions and the energy separation between the peaks are determined from this decomposition to be $\cong 0.6$ and $\cong 120\,\text{meV}$, respectively. The separation between the A and B bands in bulk CdSe is $\cong 26\,\text{meV}$ [2.13]. The larger energy separation between the A and B transitions in the quantum dot, compared with bulk, is probably due to the larger confinement of the B valence band state. In other words, the smaller effective mass of the B hole leads to a larger confinement energy and, consequently, a larger energy separation from the A transition in the dot. The observed energy separation of $\cong 446\,\text{meV}$ between the A and C^* transitions is close to both the energy separation between the $E_{1p,1p}$ and $E_{1s,1s}$ transitions and the split-off energy in CdSe ($433\,\text{meV}$).

These assignments are all in agreement with the nonlinear spectra as discussed below. The spectral width of the resonances in the linear absorption spectrum is a result of a combination of homogeneous and inhomogeneous broadening. Since the resonance energy of the quantum-confined transition varies with crystallite radius as $1/R^2$, the inhomogeneous broadening is expected to be mostly due to the crystallite size distribution.

The origin and dynamics of optical nonlinearities of the quantum confined transitions were investigated using femtosecond laser pulses. Furthermore, information about the homogeneous linewidth of the inhomogeneously broadened transitions was also obtained [2.129, 130]. We employed differential transmission spectroscopy with 60–100 fs laser pulses for these investigations. The pump

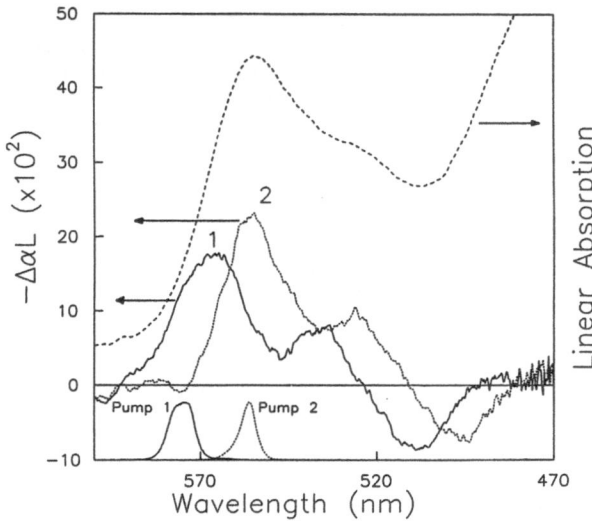

Fig. 2.20. The differential transmission of the 600°C sample for two pump wavelengths inside the A transition at 10 K. The energy positions of the pumps are indicated and the linear absorption spectrum is plotted. The linear absorption coefficient at the peak of the lowest quantum confined transition, A, is $\alpha l \sim 2.0$

pulse was tuned inside the first electronic transition, and a broad-band cross-polarized probe pulse measured the changes induced by the pump as a function of time delay between the pulses. The change in the absorption coefficient, $\Delta \alpha L$, was obtained. Figure 2.20 shows a typical result of such a measurement at low temperatures for two pump wavelengths in the A transition. Differential transmittance for the two pumping wavelengths, the energetic position of the pump, and the linear absorption spectrum of the sample are shown. The time delay between the pump and probe pulses was fixed at a value that gave the maximum signal ($t_p \cong 0.1\,\text{ps}$). The short delay for the femtosecond pump and probe pulses was also important in order to investigate mainly the intrinsic electronic effects and to reduce effects associated with traps [2.131].

It is clear from Fig. 2.20 that spectral holes may be burned at $T = 10\,\text{K}$. The spectral hole moves with the pump wavelength as expected for an inhomogeneously broadened transition. At low temperatures, the width of the spectral hole is a nonnegligible fraction of the total linewidth. At room temperature, the whole absorption is bleached, consistent with previous observations [2.132]. The width of the spectral hole increases with increasing excitation intensity. The hole width at the lowest intensities yields the homogeneous width and consequently the T_2 time. The measurement of the spectral width of the induced holes at low intensities yields a lower limit value of $25 \pm 10\,\text{fs}$ for the polarization decay time T_2. The excited state relaxation times T_1 were also measured using femtosecond four-wave mixing. Smaller dot sizes were found to lead to faster recovery times [2.129].

An interesting feature to note in Fig. 2.20 is the simultaneous bleaching of both the A and B transitions. The bleaching near the A transition is larger than that around the B transition for a given pump intensity. The bleaching behavior of A and B as a function of intensity *and* time is the same. The bleaching of the transitions is the result of the excitation of an electron to the $1s$-state which greatly reduces, simultaneously, the transitions from all hole subbands ($1s$). State filling and Coulomb interaction between injected electron-hole pairs are the responsible mechanisms for the bleaching. Generation of one electron-hole pair causes saturation of the one-pair transition and changes the excitation spectrum for the second pair. In particular, the Coulomb interaction between the pairs is responsible for the significantly different two electron-hole pair spectrum compared with the one electron-hole pair spectrum.

As shown in Fig. 2.20, there is an induced absorption feature on the high energy side of the B transition (the negative part of the signal). We rule out broadening as the origin of this induced absorption features since (i) it shifts with the burned hole when we tune the pump laser, and (ii) it appears only on the high-energy side of the transition. The origin of the increasing absorption is assigned to the generation of two-pair states in the quantum dot. The pairs are created from the sequential absorption of one pump photon and one probe photon. In other words, the creation of the first electron-hole pair in the dot by the pump-photon absorption and the resulting four-particle Coulomb interaction modifies the absorption spectrum of the probe: the one-pair transitions are bleached and new transitions become allowed. To obtain evidence for this hypothesis, we have theoretically analyzed the observed features.

The theoretical analysis of the linear and third-order nonlinear properties included the one and two electron-hole pair (EHP) states [2.43]. These investigations consistently led to the conclusion that Coulomb effects are important even for the smallest quantum dots. The "binding energy" of the energetically lowest two-pair state (loosely referred to as "biexciton"), which is a measure of the strength of the Coulomb interaction, increases with decreasing dot radius R. Using numerical matrix diagonalization techniques, Hu et al. [2.43] obtained the energies and wavefunctions for the one- and two-pair ground state and for all the excited pair states included in the basic set. With these wavefunctions, the various dipole matrix elements for transitions between ground state and the one- and two-EHP states were evaluated. Inserting the results into the equation for the two-beam third-order susceptibility yielded the changes in absorption, $\Delta\alpha$, which should be seen in a pump-probe experiment. The theoretical results for the normalized absorption changes for $R/a_0 = 1.0(0.5)$ are plotted in Fig. 2.21. The significant features are a decreasing absorption around $(\hbar\omega - E_g)/E_R \cong 6(34)$, which is the bleaching of the energetically lowest one-EHP resonance, and an increasing absorption in the region around $(\hbar\omega - E_g)/E_R \cong 10(52)$. The inset of Fig. 2.21 shows the one- and two-pair transition dipole matrix elements as dashed and full lines, respectively. The ground-state biexciton is clearly seen on the low-energy side of the lowest EHP state. The higher biexcition states which are energetically between the two lowest one-pair resonances are also seen. The induced absorption which would be expected for the ground-state biexciton is not seen due to the assumed relatively large homogeneous broadening. However, the existence of an induced absorption on the high-energy side of the one-pair resonance is clear. This feature closely resembles the experimental result of Fig. 2.20

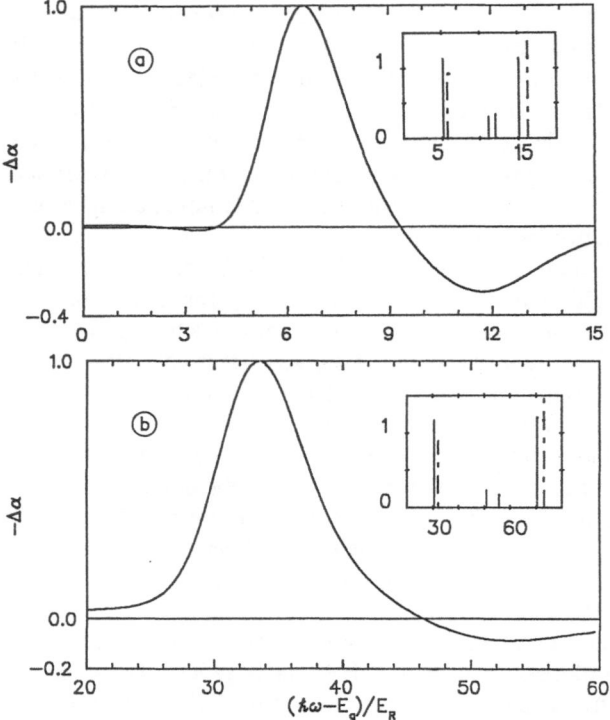

Fig. 2.21a,b. Computed absorption change $\Delta\alpha$ for semiconductor quantum dots, assuming pump-probe excitation and pumping into the energetically lowest one-pair state. The inset shows the transition dipole matrix elements for the relevant one- and two-EHP states as dashed and full lines, respectively. (a) $R/a_0 = 1$ and the broadening $\hbar\gamma/E_R = 3$. (b) $R/a_0 = 0.5$ and $\hbar\gamma/E_R = 9$

and is due to the excitation of two-pair states corresponding to two electrons in the 1s-state, and two holes in the 1p-state, or one hole in the 1s-state and the other hole in the 2s-state, respectively. These transitions would be dipole forbidden without inclusion of the Coulomb interaction. Furthermore, they do not exist in bulk semiconductors since the corresponding biexciton states are not energetically bound.

The recovery of the nonlinear spectra was measured as well. Figure 2.22 shows the nonlinear behavior at different time delays between the pump and probe pulses. The data were taken at 10 K for the 600°C sample. At least a two-component decay is observed, a fast decay followed by a slower one. About 50% of the bleaching recovers in 50 ps. The remaining part of the bleaching lasts for times which are longer than our 500 ps delay line. The fast component of the decay is attributed to the carrier recombination, eventually, through surface states. The slow component of the decay is attributed to the recombination involving trapped states or impurities. It has been suggested that selenium anion vacancies, i.e. Cd^{2+}, which are the most likely defects in these and similar samples, are the possible traps [2.133].

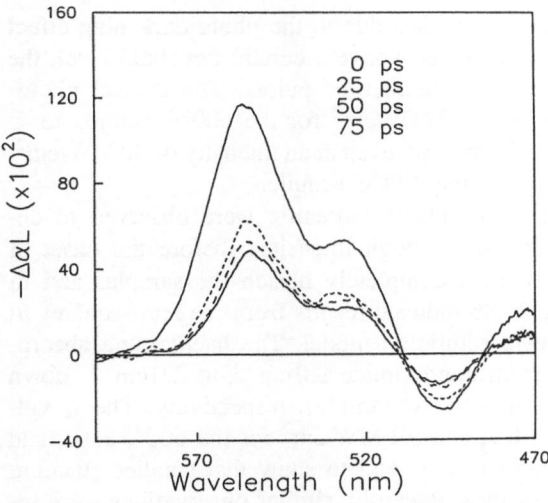

Fig. 2.22. The differential transmission for the 600°C sample at $T = 10\,\mathrm{K}$ for various pump-probe delays, showing the carrier decay dynamics

The quantum dotsize-dependence of the optical nonlinearities for three samples was also investigated [2.134, 135] using a self-saturation experiment with 3-ns pulses. The absorption versus intensity for the samples, which clearly exhibit bleaching, is shown in Fig. 2.23a for room temperature and in Fig. 2.23b at 10 K.

These data were fit to the simple saturable absorption model with a nonsaturating background absorption

$$\alpha = \frac{\alpha_0}{1 + I/I_s} + \alpha_B , \qquad (2.16)$$

where α_0 and α_B are the saturable and nonsaturable (background) absorption, respectively. For the room temperature data in Fig. 2.23a, however, we are unable to obtain the saturation intensity I_s and the background absorption α_B because of

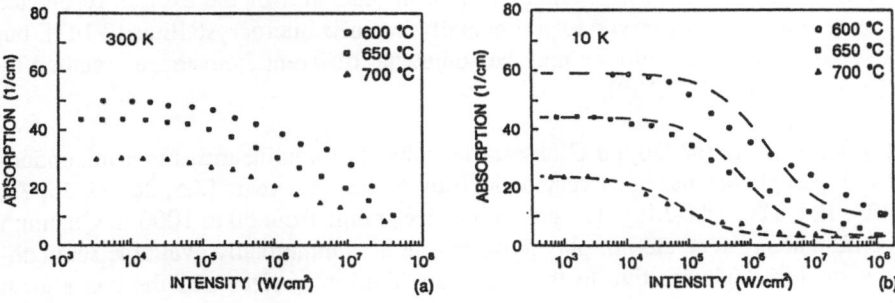

Fig. 2.23. Single-beam absorption saturation data taken at (a) 300 K; (b) 10 K for the three samples

the lack of sufficient data at higher intensites due to the photo-darkening effect [2.136, 137]. When the irradiation increased above a certain threshold level, the transmission decreased with time, i.e., number of pulses. The critical photo-darkening intensity increased from $< 20 \, \text{MW/cm}^2$ for the 600°C sample to $< 35 \, \text{MW/cm}^2$ for the 650°C sample. Moreover, even at an intensity of $30 \, \text{MW/cm}^2$, photo-darkening was not observed for the 700°C sample.

At low temperature, the effects of photo-darkening were observed to decrease drastically. Taking data at large enough intensites (before the onset of photo-darkening) allowed us to almost completely bleach the samples and fit the data; the dashed lines in Fig. 2.23b indicate results from the least-squares fit using the aforementioned two-level saturation model. The background absorption α_B was found to decrease from approximate $9.0 \, \text{cm}^{-1}$ to $2.0 \, \text{cm}^{-1}$, down to $2.7 \, \text{cm}^{-1}$ for the 600°, 650°, and 700°C samples, respectively. The I_s values obtained from the fit are 1.9, 0.7, and $0.1 \, \text{MW/cm}^2$ for the 600°, 650°, and 700°C samples, respectively. These results tend to show that smaller quantum dots exhibit larger saturation intensities. Recently, similar observations were reported by *Hall* and *Borrelli* [2.138] in these quantum-confined glasses at room temperature using a picosecond laser system. More specifically, in both experiments a laser pulse energy of $\cong 1.5 \, \text{mJ}$ produced a similar transmission change of $\Delta T \cong 0.03$ at the excitation wavelength of $4\lambda \cong 580 \, \text{nm}$. Considering the fact that our experiment was performed with 3-ns pulses while *Hall* and *Borrelli* [2.138] used a-few-picosecond pulses, one may conclude that the effective response time of the nonlinearity must be longer than a few nanoseconds (in order for the transmission changes to be the same for the same laser energy). This fairly long response time is again attributed to the presence of impurities or traps. This observation indicates that, at least in the picosecond and nanosecond time domains, the nonlinearity is strongly affected by impurities or traps.

Before we leave this subject, it should be mentioned that further theoretical work on optical nonlinearities in various quantum-confinement regimes has been discussed by several groups [2.8, 40, 42, 139–142]. For large quantum dots whose sizes considerably exceed the bulk exciton diameter, an excitation-induced blue shift of the exciton resonance has been predicted as a consequence of the plasma screening of the attractive Coulomb interaction between electrons and holes [2.139]. In fact, an excitation-induced blue shift of the exciton resonance has recently been observed experimentally in CuBr microcrystallites [2.143], but the experimental conditions may be somewhat different from those assumed by *Banyai* and *Koch* [2.139].

(c) **Semiconductor Doped Glasses.** The subject of nonlinearities in semiconductor doped glasses has been very popular in the last few years [2.5, 26, 33, 75, 76, 123, 124, 126, 144–149]. The crystallite sizes range from 50 to 1000 Å. Quantum confinement effects do not play a major role in commercially available semiconductor doped glasses due to the broad distribution in sizes and the consequent large inhomogeneous broadening, which washes out these effects [2.31].

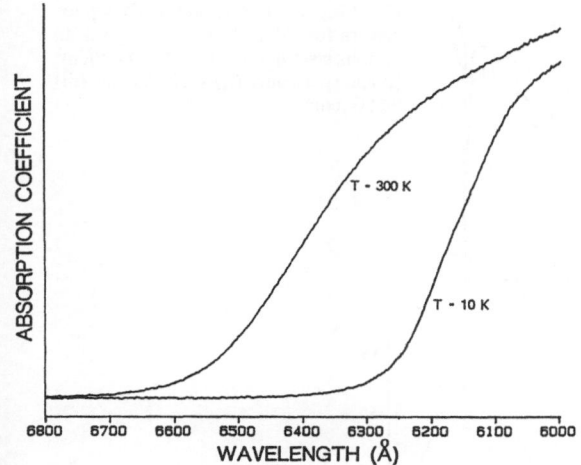

Fig. 2.24. Absorption spectra of a typical commercial semiconductor-doped glass at room and liquid helium temperature

These materials have been available commercially for many years as sharp-cut color filters, manufactured by Corning and Schott, for use as linear blocking elements. Figure 2.24 shows the absorption spectra in a typical commercial color glass at room and liquid-helium temperatures. The spectra do not show any sharp structures even at low temperatures, suggesting a large inhomogeneous broadening effect. The spectra display the band-edge behavior of the semiconductor. The importance of semiconductor-doped glasses for nonlinear optical applications was realized first by researchers at Hughes Laboratories in 1983, who reported degenerate four-wave mixing in these glasses [2.123]. During the past few years, the origin of the nonlinearity [2.33, 75], the magnitude of the nonlinear index at various wavelengths in the vicinity of the band-gap [2.75], and the femtosecond dynamics of such glasses have been investigated [2.76, 126, 146]. Phase conjugation, four-wave mixing [2.33, 145, 150, 151] and luminescence [2.147, 148, 150, 152] studies were reported. The use of semiconductor-doped glasses as nonlinear waveguides has also been proposed [2.150, 153] and has become an attractive challenge for researchers in this field [2.154–156]. A detailed review of these results is beyond the scope of this book, so we only summarize some of the findings.

The physical origin of the optical nonlinearities has been studied using a two-beam pump-probe experiment [2.75]. The transmission of a weak probe beam is monitored as a function of wavelength for various pump intensities. The pump-induced changes in the probe transmission spectra at room temperature are plotted in Fig. 2.25a for various pump intensities. The spectrum labeled iii) corresponds to the highest pump intensity, while the spectrum labeled i) is the linear absorption spectrum. With increasing incident intensity, bleaching of the band-edge absorption spectrum is observed, which is mainly due to band filling. The change in absorption coefficient $\Delta\alpha(\omega)$ obtained by subtracting spectrum i) from iii), i.e., $\Delta\alpha(\omega) = \alpha(\omega, I = 3\,\mathrm{MW/cm^2}) - \alpha(\omega, I = 200\,\mathrm{kW/cm^2})$, is displayed in Fig. 2.26a by the solid curve. The corresponding change of the refractive index is obtained by a Kramers-Kronig transformation.

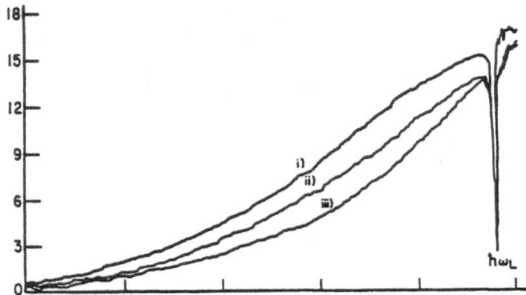

Fig. 2.25. Room-temperature absorption spectra for $CdS_{0.9}Se_{0.1}$-doped glass at various laser intensities. (i) 200 kW/cm^2 (linear spectrum), (ii) 1.2 MW/cm^2, (iii) 3 MW/cm^2

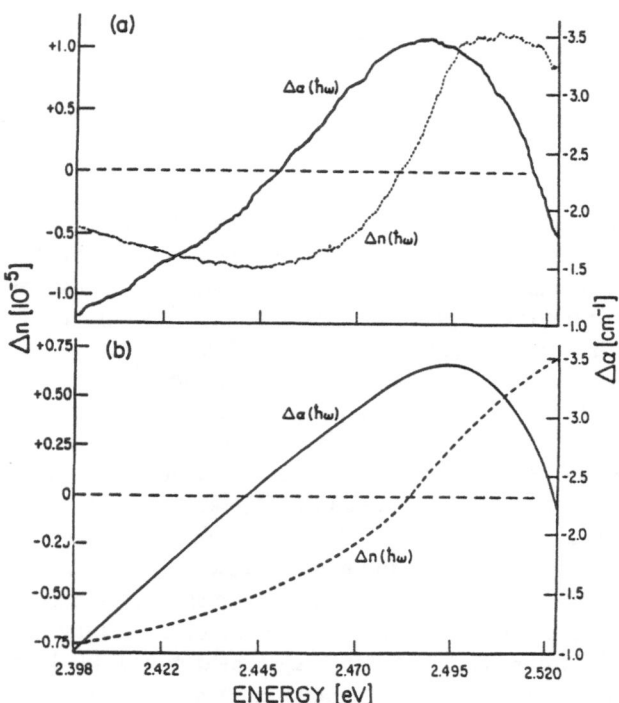

Fig. 2.26a,b. Absorptive changes and corresponding dispersive changes of $CdS_{0.9}Se_{0.1}$-doped glass. (a) Experimental results: $\Delta\alpha = \alpha(I = 3\,\mathrm{MW/cm^2}) - \alpha(I = 200\,\mathrm{kW/cm^2})$. (b) Theoretical results: $\Delta\alpha = \alpha(N = 10^{18}\,\mathrm{cm^{-3}}) - \alpha(N = 0)$

The experimental results have been compared with the plasma theory [2.53] discussed earlier in this review. Additionally, the Maxwell-Garnet equation [2.157] is used to express the average dielectric function ε^{av} of the composite material (glass plus embedded semiconductor crystallites),

$$\varepsilon^{av}(\omega) = \varepsilon_g \frac{\varepsilon(\omega)(1 + 2p) + 2\varepsilon_g(1 - p)}{\varepsilon(\omega)(1 - p) + \varepsilon_g(2 + p)} , \qquad (2.17)$$

where $\varepsilon_g = 2.25$ is the dielectric constant of the host glass, p is the fraction of the total volume occupied by semiconductor crystallites, and $\varepsilon(\omega)$ is the complex

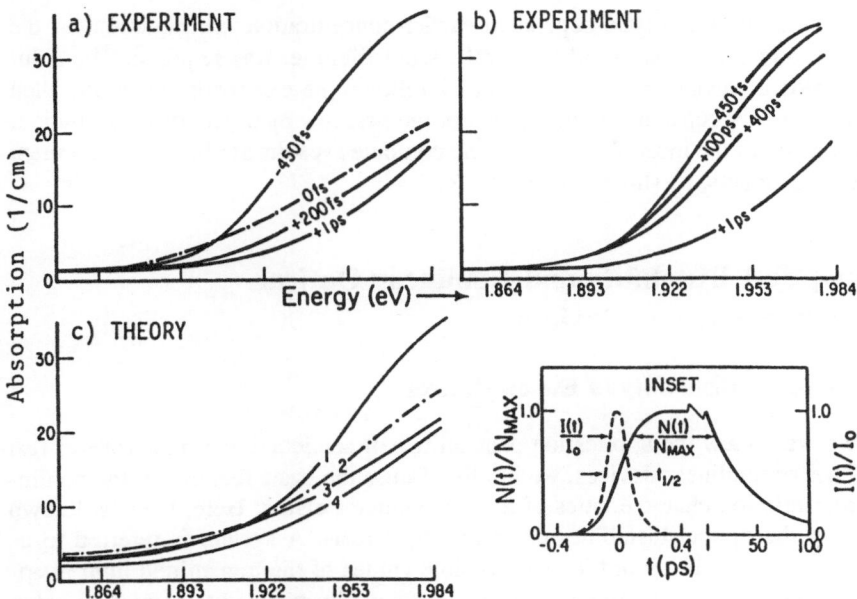

Fig. 2.27. (a,b) Room-temperature absorption spectra of a 750-μm thick glass doped with $CdS_{0.2}Se_{0.8}$ microcrystallites for various pump-probe delays. (c) Plasma theory calculation: 1) $T = 300\,\mathrm{K}$, $N = 10^{15}\,\mathrm{cm}^{-3}$; 2) $T = 550\,\mathrm{K}$, $N = 5 \times 10^{17}\,\mathrm{cm}^{-3}$; 3) $T = 450\,\mathrm{K}$, $N = 9.5 \times 10^{17}\,\mathrm{cm}^{-3}$; 4) $T = 350\,\mathrm{K}$, $N = 10^{18}\,\mathrm{cm}^{-3}$. *Inset*: Relative temporal evolution of the excitation pulse, $I(t)$, and the carrier concentration $N(t)$, $N_{\mathrm{max}} = 0.006 N_0$

dielectric function of the semiconductor crystallites. The detailed comparison between theory and experiment is shown in Fig. 2.26b, where we plot the theoretical and experimental changes in the absorption and the corresponding refractive indices $\Delta\alpha(\omega)$ and $\Delta n(\omega)$, respectively.

The dynamics of the optical nonlinearities in these materials have also been investigated using femtosecond light pulses [2.76, 126, 146]. Figure 2.27a,b displays absorption of the probe as a function of wavelength for various time delays between pump and probe. The spectrum-labeled $-450\,\mathrm{fs}$ is taken when the probe precedes the pump and therefore is representative of the unexcited semiconductor, showing the band-edge absorption of the microcrystallites. At later times, a saturation of the absorption just above the band edge is observed. This suggests that carriers initially created with a few LO-phonon energies above the band-gap have relaxed to the bottom of the band and initiated the band-filling mechanism. After a few picoseconds, the saturation starts to recover, indicating the onset of electron-hole recombination, with an almost complete recovery in $\cong 50\,\mathrm{ps}$. However, a small portion of absorption reduction does not fully recover and persists for times in excess of $500\,\mathrm{ps}$. This may be attributed to carriers confined to traps near the band edge.

These experimental results were also compared with the plasma theory [2.53]. For this comparison, a knowledge of the injected carrier density, which is obtained by solving the time-dependent carrier-concentration rate equation for the femtosecond excitation and the recombination lifetime, was required. The solution yields a carrier density that initially follows the excitation pulse and then becomes nearly constant for an interval of the first few picoseconds of excitation, as displayed in the inset of Fig. 2.27. The computed results are in good agreement with experiments, as shown in Fig. 2.27c.

2.3 Optical Bistability and Nonlinear Optical Semiconductor Devices

2.3.1 Optical Bistability in Etalon Devices

Before we review the application possibilities of semiconductor materials as fast switches or nonlinear devices, we briefly discuss the basic features of the nonlinear transmission characteristics of a semiconductor etalon. Here, the best-known effect is the optical bistability or optical hysteresis. A system is referred to as optically bistable if it exhibits two possible values of the transmitted light intensity for one value of the input intensity. The actually realized transmission value depends on the excitation history. The system is in a different state depending on whether the input intensity is increased from zero or decreased from a sufficiently high level.

To model the effect of optical bistability, it is most convenient to discuss the situation where the nonlinear medium is placed inside a Fabry-Perot etalon. The transmission of such a system is determined by the equation [2.158–163]

$$I_t = I_0 T^2 \frac{1}{(e^{\alpha(\omega,N)L/2} - Re^{-\alpha(\omega,N)L/2})^2 + 4R\sin^2[\delta + \omega\Delta n(\omega, N)L/c]} , \quad (2.18)$$

where I_0 and I_t are the incident and transmitted light intensities, respectively [2.164]. The reflection coefficient of the mirrors is denoted by R and the transmission coefficient by $T = 1 - R$. In many practical applications, these mirrors are actually the endfaces of the semiconductor crystal itself, and R is just the natural reflectivity. R can be increased through additional high-reflectivity coatings evaporated onto these surfaces. The quantity δ is the linear round-trip phase shift of the light in the resonator (initial detuning), L is the length of the medium, and $\alpha(\omega, N)$ and $\Delta n(\omega, N)$ are the carrier-density-dependent absorption and refractive-index change, respectively. The Fabry-Perot transmission depends on the carrier density N through α and Δn. The carrier density in turn is determined by the intensity I of the light inside the resonator through a rate equation of the form

$$\frac{dN}{dt} = -\frac{N}{\tau} + \frac{\alpha(\omega, N)}{\hbar\omega}I_0 . \quad (2.19)$$

The incident intensity I_0 may vary in time, as in the case of an excitation pulse,

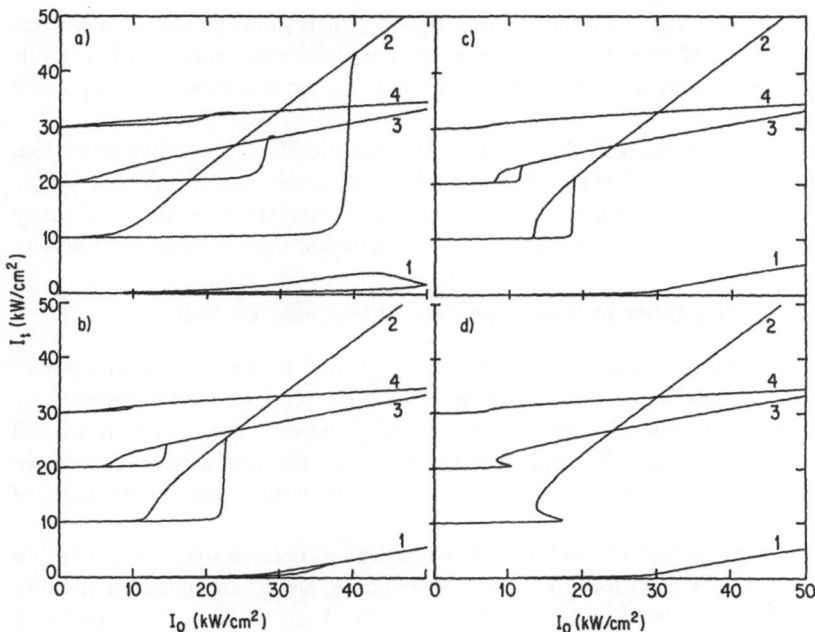

Fig. 2.28a-d. Transmitted intensity versus input intensity computed for a $\cong 2\,\mu$m GaAs etalon at room temperature for the excitation energy $\hbar\omega = 1.4032\,$eV well below the exciton resonance at $1.420\,$eV. Parts (**a–c**) are obtained assuming pulsed excitation with a triangular pulse of full width 10τ (a), 100τ (b), and 1000τ (c). Part (**d**) shows the steady-state results. The different curves 1–4 in each figure are for the detunings $\Delta\hbar\omega = \hbar\omega_{R} - \hbar\omega = -0.0170\,$eV, $-0.0142\,$eV, $-0.0115\,$eV, and $-0.008\,$eV, respectively. $\hbar\omega_{R}$ is the resonator eigenfrequency, τ is the carrier lifetime, and the mirror reflectivity $R = 0.9$. The baseline for the transmitted intensity in curves 2, 3, and 4 has been shifted by 10, 20, 30 kW/cm^2, respectively

but the temporal variation has to be slow on the time scale of the resonator round-trip time for steady-state operation.

The coupled equations for carrier density and Fabry-Perot transmission have been solved numerically by *Warren* et al. [2.165] for the example of a GaAs resonator at room temperature using $\alpha(\omega, N)$ and $\Delta n(\omega, N)$ computed from the plasma theory discussed earlier. Examples of the results are shown in Fig. 2.28a–c for different temporal widths of the exciting pulse and in Fig. 2.28d for the steady state. The curves 1–4 in these figures are obtained for varying resonator lengths L, causing different detunings of the excitation frequency with respect to the nearest resonator eigenfrequency. The curves 2 and 3 show well-developed hysteresis loops similar to those observed in experiments. The size of these loops increases if the temporal pulse width decreases. This is the well-known effect of dynamical hysteresis. These dynamic effects can even produce seemingly bistable loops, as in curves 1 and 4, which vanish for longer pulses.

Optical bistability of electronic origin has been observed in many semiconductor materials, such as bulk GaAs and GaAs multiple quantum wells [2.166–

168], InSb [2.2], CdS [2.60] and many others (for a more complete reference list see [2.4]). Additionally, there have been various observations of bistability caused by thermal nonlinearities, which will not be discussed in the present chapter.

It has to be emphasized that under a slightly modified operating condition, namely at a slightly different initial detuning, the same bistable device shows thresholding or limiting characteristics with no hysteresis behavior. For many applications the bistability may not be needed and just thresholding is sufficient.

2.3.2 Optical Bistability in Semiconductor Waveguide Devices

In contrast to etalon devices, the light travels parallel to the surface in a waveguide device. There are several possible designs for semiconductor waveguides, such as planar guides, channel structures, ridge waveguides and strip-loaded structures, to name only a few examples [2.169–172]. The merits of the respective designs depend to a large extent on the desired operation and on the ease of fabrication.

Most of the reported bistability in waveguides has been of thermal origin. For the case of a MQW strip-loaded channel waveguide, electronic optical bistability has been reported recently [2.173]. The guide consisted of an AlGaAs cladding layer on top of a MQW structure. A ridge was edged into the cladding layer to provide transverse confinement of the light travelling in the MQW structure via the difference of the effective refractive indices (effective index method [2.170]). The power coupled into the waveguide was estimated to be 30 mW [2.173] and up to two bistability hysteresis loops were observed, indicating that phase shifts in excess of 2π are attainable. This result shows that bistable waveguide devices in MQW GaAs can be made to operate but, at least until now, the required power is higher than in etalons. The higher power does not seem to be too surprising, however, considering the fact that in addition to the losses which are present in waveguide and resonator structures, the light in the waveguides experiences guiding losses through evanescent tails, imperfect surfaces, and other irregularities in the spatially quite extended structures.

In principle, semiconductor waveguide devices can perform useful operations which are not possible in etalon structures [2.154, 174–177]. One example is the so-called nonlinear directional coupler consisting of two closely spaced parallel guides. The light is inserted into only one of the guides, but it tunnels into the neighboring guide after a characteristic length, which is called the beat length. The beat length is a very sensitive function not only of the design parameters like the separation of the guides but also of the refractive indices of the guides. Since the refractive index changes with changing carrier density, the beat length changes too. Hence, for a directional coupler which has just the linear beat length, the light injected into one guide is totally coupled into the other guide for low light intensity, whereas it may stay in the original guide for sufficiently high intensities. Such a nonlinear directional coupler has been realized [2.174] in a metal-strip loaded GaAs structure, where transverse light-confinement and probably also carrier-confinement is introduced by the strain field at the edges of the metal strips. Nonlinear directional coupling using 10-ps pulses has also been realized recently in single-mode GaAs/AlGaAs strip-loaded waveguides [2.176].

The advantages of the waveguide devices lie in the fact that, in principle, large light intensities may be kept in the devices for relatively long distances. In addition, they may perform operations such as nonlinear directional coupling for routing and switching purposes which cannot easily be simulated by other devices such as etalons. However, comparison between etalon and waveguide devices based on semiconductor band-edge nonlinearities for similar operations like optical bistability and switching, reveals that etalons have better performance characteristics. For example, the transmission throughput of GaAs etalons is much better than has been demonstrated so far in waveguides. This may be due to the fact that waveguide devices suffer from guiding and coupling losses in addition to the intrinsic material losses (the intrinsic material loss problem is the same for both etalons and waveguides). The lack of semiconductor waveguides with high throughput (more than 10%) and with electronic nonlinearity emphasizes that these devices are currently limited by the technology of fabricating low loss waveguide devices.

2.3.3 Optical Bistability in SEEDs

Self-electro-optic-effect devices (SEEDs) [2.90] are cavityless devices that operate on the basis of increasing absorption. Light absorption causes an electro-optic shift of the band edge through electrical feedback. Initially, an electric field is applied across the device. The absorption of the system is governed by that above the low-energy shifted exciton (due to the E-field). As the light intensity increases, the flow of current through the device as a result of the generated electron-hole pairs causes a drop in the applied E-field. The exciton shifts back toward its unbiased position and the device transmission decreases. The absorption of the system in the "off" state is then governed by that of the exciton peak. Quite clearly then, SEEDs do not work in bulk GaAs at room temperature; the existence of a well-defined exciton resonance is essential. SEEDs are hybrid devices, so their capacitance can be increased to trade off speed for reduced power: 400 ns at 3.7 mW or 1.5 ms at 670 nW with similar energies. A SEED's contrast depends upon the above-mentioned absorption difference and therefore can be enhanced by using devices with thicker active media, whereas its transmission depends upon the lower of the two absorptions. Thus there is also a trade off between bistability contrast and insertion loss. The advantages of the SEEDs are their interface to electronics, the absence of an external feedback and their speed-power tunability. Large device size and high absorption (even in the "on" state) are the disadvantages of this type of device.

2.3.4 Thermal Etalon Devices

ZnS and ZnSe thin-film interference filters with thermal nonlinearities have been investigated for their potential application in optical signal processing and optical computing. Considerable research has been devoted to their basic properties such as the nonlinear mechanism, switching powers, switching times, filter design, and crosstalk [2.88, 89, 178] since the first observation of optical bistability in these filters by *Karpushko* and *Sinitsyn* [2.179]. Cascading, gain, and a triple-bistable-element loop have been demonstrated [2.183, 184], and parallel operation of multiple "pixels" (logic gates) has also been attempted. The observation of simultaneous bistable switching of multiple pixels on ZnS and ZnSe interference filters has been reported, the recognition of a simple pattern has been demonstrated [2.178], and simple symbolic substitution has been performed [2.182, 183].

2.3.5 Optical Logic Gates

In addition to the one-wavelength optical bistability, two-wavelength operation of nonlinear etalons has demonstrated various optical logic operations such as AND, OR, and NOR. The principle of such operations is the shifting of the Fabry-Perot transmission peak in response to input pulses. The nonlinear medium must be such that the absorption of one input pulse changes the refractive index at the probe wavelength enough to shift the transmission peak of the etalon by about one instrument width [2.184]. For example, if the etalon is initially tuned to the probe wavelength, a NOR-gate operation results if the control pulse is able to detune the etalon from the probe wavelength. The speed of the gate can be very high, with the potential for picosecond decision making at a gigahertz repetition rate [2.4]. Two-dimensional arrays of GaAs pixels have been demonstrated [2.185, 186].

Optical gating in GaAs has progressed considerably in recent years. A 1 ps switch-on time of a GaAs MQW gate has been demonstrated [2.187]. This time indicates how fast the etalon peak shifts in response to an input pulse with an above-band-gap frequency. Recovery of the gate requires removal of the carriers produced by the input pulse. This recovery takes more than 10 ns in the usual GaAs and MQW etalons because of the long lifetime of carriers. The recovery time has been shortened to 200 ps [2.188] (detector limited) using surface recombination in an etalon with no AlGaAs outside layers (the AlGaAs windows are normally used to stop etching of the GaAs substrate). A 30-ps recovery and 70-ps cycling time have also been demonstrated in similar etalons [2.189]. *Ojima* et al. [2.190] have observed MQW NOR-gate operation using diode lasers as the only light sources.

2.3.6 Use of the Optical Stark Effect in Logic Gates

If the input-pulse frequency is tuned below the excitaton into the transparency region of the material, then the electric field of the laser may be used to switch

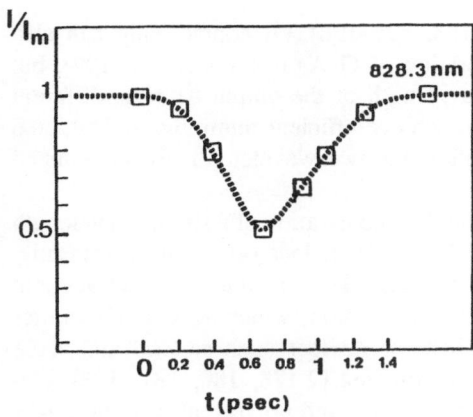

Fig. 2.29. Picosecond switching of a GaAs logic gate. The optical Stark effect has been employed to obtain these switching speeds

the etalon. The advantage of this optical-Stark-effect switching is that the carrier lifetime is no longer a limitation, as carriers are not really excited in this scheme. The disadvantage of such high speed devices is, of course, the higher input powers required to realize them. A 1-ps recovery using this effect has been demonstrated, as shown in Fig. 2.29 [2.191], which shows NOR-gates switch-on and switch-off times of $\cong 1\,\mathrm{ps}$. The advantages of nonlinear etalons are their small size, availability of large two-dimensional arrays of etalons and their high speed. Their disadvantages are the lack of picosecond gain (this will be discussed below) and large fan out. The switching energies of nonlinear etalons compete very well with other devices. The lowest demonstrated switching energy per gate is $\cong 0.6\,\mathrm{pJ}$ [2.186, 192].

An important aspect of an optical logic gate is the fan-out capability, which is essential for the cascadability needed in many practical applications. In order to achieve cascading and fan-out, the devices need to provide gain. One such device was recently demonstrated [2.193]. This device is a diode laser amplifier which is electrically biased near threshold. The device may be switched by application of a low power laser beam. Large gain values were reported. The total switching power (sum of electrical and optical power) of the device is similar to other devices.

2.3.7 Gain and Cascading Issues in Nonlinear Etalons

Some of the important requirements for optical bistable devices or nonlinear switches are cascadability, gain (or fan-out) and good contrast. These requirements are essential because in order to cascade, signal levels have to be restored after each operation in order to drive the next stage. For a reasonable fan-out, the output of one logic gate has to be used to switch several logic gates. The contrast should be high enough that the signal-to-noise ratio can be kept high throughout the system.

Single-wavelength gain and cascading have already been demonstrated for some materials, such as ZnS, ZnSe, InSb, allowing some simple digital optical circuits to be built [2.178, 180, 181, 183, 194–196]. Of course, gain can also be obtained using two-wavelength operation of GaAs devices [2.186, 192], but then cascading is difficult since the wavelength of the output beam is different from that of the input beam in order to achieve efficient pumping, and isolated resonances are not easily available to allow the two wavelengths to be switched interchangeably.

Here, we only mention the experimental demonstration of a single-wavelength cascading circuit using two 58-Å MQW bistable etalons operating in transmission mode [2.197]. With milliwatt power, these devices were operated at room temperature using pulses of about $1\,\mu W$ in duration, which was much shorter than that used previously for thermal bistable devices such as ZnS and ZnSe interference filters for demonstration experiments [2.178, 180, 181, 183]. Despite the fact that the etalons were not optimized, and no special procedure was used to stabilize the laser power, a gain of 4 was still obtained. Large contrast was also observed, with the largest contrast of $\cong 10$ for an etalon operating in transmission. Thus the requirements mentioned earlier are satisfied for microsecond pulses. Numerical simulations are also carried out for shorter pulses, and the results show that the differential energy gain vanishes for pulse durations approaching the carrier lifetime, indicating that a reduction of the carrier lifetime may be necessary to achieve single-wavelength cascading on a picosecond time scale.

For a nonlinear etalon operating in a bistable or thresholding mode (Fig. 2.30) the gain and the contrast can be defined in terms of the power of the light pulses as follows. Suppose a weak signal beam is amplified through a nonlinear etalon by a strong holding beam. We define the device gain as

$$G_{\mathrm{d}} = \frac{P_{\mathrm{u}} - P_{\mathrm{l}}}{P_{\mathrm{s}}}, \qquad (2.20)$$

where P_{u} is the output power when the device is in the upper bistable state, P_{l} is the output in the lower state, and P_{s} is the signal-beam power as shown in Fig. 2.30. This definition of G_{d} is suitable for the steady-state operation of a device, i.e. when the input pulse length is much longer than the device response

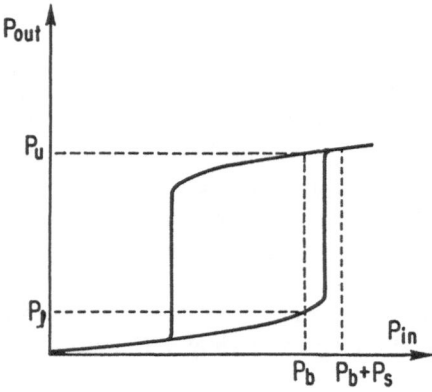

Fig. 2.30. A bistable device as a latching logic gate. P_{b} is the bias power, P_{s} is the signal power, P_{u} and P_{l} are the upper- and lower-state output powers, respectively

time. Note that G_d is limited by practical considerations such as the stability of the laser amplitude and frequency, the etalon temperature stability, etc. Theoretically, G_d can be made to approach infinity by decreasing P_s, but the switching time may increase due to the critical-slowing-down effect.

The contrast of a logic gate is defined as

$$C_d = \frac{P_u}{P_l} . \tag{2.21}$$

For most applications, the contrast represents the signal-to-noise ratio of a device. Thus a large contrast is generally desired, especially for such applications as the nonlinear decision making in an optical associative memory [2.198]. For an etalon operating in the transient mode, where the pulse length is comparable to the device response time, we can still define the device gain and the contrast in a similar way, with the powers in the above expressions replaced by energies.

The minimum gain required for cascading is $G_d > 1$. Generally, for a gate to have a fan-out of N, $G_d > N$ is needed if the output of the first gate is to be able to switch N successive gates by itself. This requirement can be reduced if the outputs of several gates are used to switch one gate (fan-in). The contrast must also be high enough that $P_u/N > R_s$ but P_l/N is too small to switch the device. This requirement in effect determines G_d. The N needed depends upon the architecture. If the simplest symbolic-substitution logic is to be implemented, $N > 2$ will be required [2.199].

GaAs nonlinear etalons can be operated in the transmission mode. Figure 2.31 shows the experimental setup for cascading with two etalons working in transmission as OR-gates where the transmitted signal of the first etalon (E1) is used to switch the second one (E2). A typical result of the cascading experiment is shown in Fig. 2.32. Note that the results shown here are for the cases where the devices are operated in the bistable mode. It is also possible to adjust the experimental parameters (such as the etalon detuning) to make the bistable loop extremely narrow to approach the case of thresholding operation, which was also experimentally demonstrated. The bias-beam powers in these experiments were

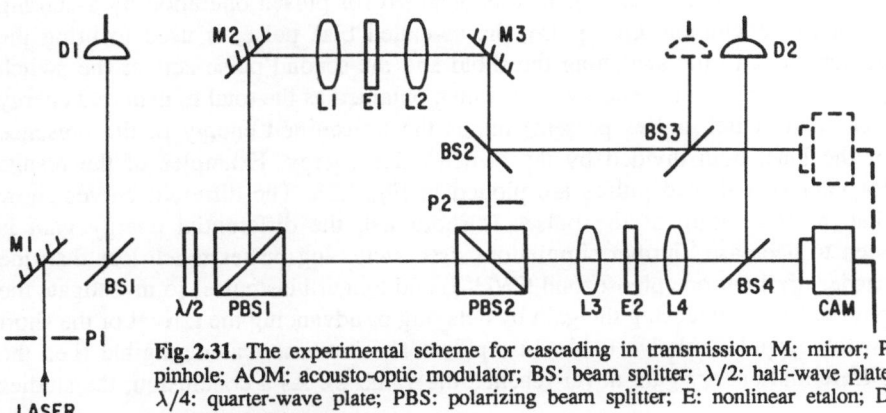

Fig. 2.31. The experimental scheme for cascading in transmission. M: mirror; P: pinhole; AOM: acousto-optic modulator; BS: beam splitter; $\lambda/2$: half-wave plate; $\lambda/4$: quarter-wave plate; PBS: polarizing beam splitter; E: nonlinear etalon; D: detector; CAM: camera

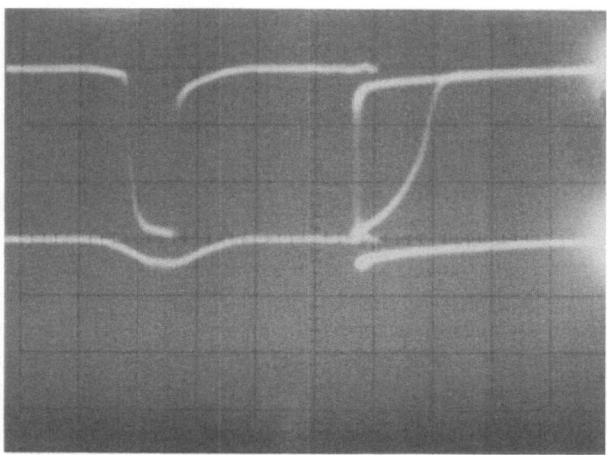

Fig. 2.32. Results of the cascading experiment corresponding to Fig. 2.31. The traces on the left are the ouput vs input plots, and the traces on the right are the outputs from the second stage. The upper traces represent the case when the first stage is "off", and the lower traces represent the case when the first stage is "on". The total input peak-power for each device is about 30 mW, and the input pulse length is 1.5 μs

20–40 mW in order to observe sufficient gain, and the etalon which had a lower threshold was chosen to be used in the first stage. The largest gain observed was 4, where a 0.25 mW change of the output signal of the first stage induced a change of 1 mW in the output from the second one. The contrast at these power levels was 5–8; it could be increased to about 10 with larger input powers (45–50 mW).

To study the variation of the differential energy gain as a function of the temporal pulse width, numerical simulations of all-optical room-temperature GaAs devices were performed. In these simulations, the transmission of a GaAs Fabry-Perot resonator was computed using the procedure given in Sect. 2.3.1, with the absorption coefficient $\alpha(\omega, N)$ and the nonlinear part of the refractive index $\Delta n(\omega, N)$ being obtained from the plasma theory.

The differential energy gain was obtained for pulsed operation by assuming two-pulse excitation. One pulse, the so-called bias pulse, is used to bring the device close to the switching threshold and the second pulse acts as the switch pulse. The differential energy gain is then obtained as the total transmitted energy (bias and switch pulses present) minus the transmitted energy in the presence of the bias, both divided by the switch-pulse energy. Examples of the results for Gaussian-shaped pulses are plotted in Fig. 2.33. The different curves show that, as the length of the pulses is shortened, the differential energy gain is seen to decrease. Further simulations used switching pulses much less than the carrier lifetime (one picosecond FWHM) and longer bias pulses to investigate the possibility of increasing the gain by delaying or advancing the arrival of the short switching pulse relative to the bias pulse. The increase was negligible (i.e., the differential energy gain never reached the value 2). As a conclusion, the studies

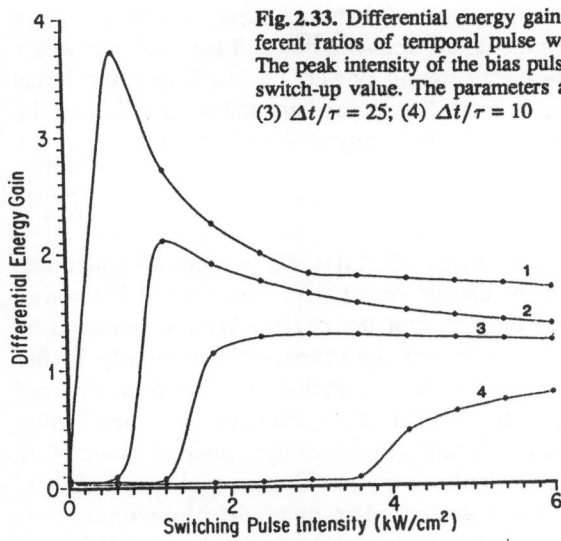

Fig. 2.33. Differential energy gain versus switch-pulse intensity for different ratios of temporal pulse width Δt to carrier relaxation time τ. The peak intensity of the bias pulse was kept just below the steady state switch-up value. The parameters are (1) $\Delta t/\tau = 100$; (2) $\Delta t/\tau = 50$; (3) $\Delta t/\tau = 25$; (4) $\Delta t/\tau = 10$

indicate that passive nonlinear etalon devices cannot be expected to exhibit useful differential energy gain for operating pulses shorter than roughly 10τ where τ is the carrier lifetime. In order to achieve single-wavelength, cascadable optical logic gates that can operate with picosecond pulses, it will be necessary to reduce the carrier lifetimes in the semiconductor material.

Since the bistable behavior of GaAs MQW etalons is very similar to that of bulk GaAs etalons, one can predict that the above conclusions hold at least qualitatively for the GaAs MQW etalons used in the experiment. Indeed, we observed a decrease of the gain as the pulse length is shortened to several hundred nanoseconds. As described earlier, some efforts to reduce the carrier lifetime have already been made, such as the use of thin samples without protective AlGaAs "windows" [2.188] and increasing surface recombination rates by etching of surface features [2.188]. The reduction of pixel size also serves this purpose [2.186, 192]. Another promising approach could be the use of semiconductor microcrystallites. For the case of II-VI compounds, it has been shown that the lifetime is reduced by two to three orders of magnitude on reducing the crystal from bulk to microcrystallites on the order of around 100 Å [2.76, 146].

2.4 Conclusion

Semiconductors offer a variety of optical nonlinearities that can be employed for optical computing, optical signal processing and neural network implementations. All the optical nonlinear devices and logic gates that have been described in this chapter, irrespective of whether they are hybrid or all-optical, follow some common rules. For example, there is a trade-off between speed and the operating power of the device and between the device transmission and the nonlinear index change (which in turn controls the operating power, i.e. the larger the index change, the smaller the required power). These fundamental trade-offs may be understood simply from (2.19) and (2.1). In steady state (2.19) reads

$$N = \alpha(\omega, N) I_0 \frac{\tau}{\hbar \omega} . \tag{2.22}$$

For switching applications, a certain value of $\Delta n(w, N)$ is required which can be obtained with a given value of carrier density N. Equation (2.22) shows that if one reduces τ (increasing the speed of the device), then I_0 needs to be increased in order to keep N constant (power–speed trade-off). Similarly, higher device transmission is possible by moving the operating wavelength away from the exciton peak so that the tail absorption decreases. However, detuning farther from the exciton also reduces $\alpha(\omega, N)$, and consequently a smaller value of N is reached which leads to a smaller nonlinearity and higher required power.

Different applications of nonlinear devices place quite different requirements on the bistable etalons. For example, the symbolic-substitution logic [2.178, 183, 199] requires that large arrays of pixels (probably $> 100 \times 100$) be operated simultaneously, which means that good uniformity in a relatively large etalon area (on the order of $1 \, cm^2$) is needed. In some applications, one may be able to reduce the speed requirement for a single gate in exchange for a lower operating power. High throughput may be achieved using simple interconnection patterns and low fan-out requirements in this scheme. Applications along this line also include spatial-light-modulators with sub-microsecond (single-wavelength, cascadable) or even picosecond (two-wavelength) addressing times. Another class of application is high-speed sequential processing, which takes advantage of the fast switching speed of a single device, like the all-optical compare-and-exchange switch [2.200]. With GaAs optical logic gates, the data rate of a single self-routing channel can be 1–10 Gbits/s (assuming 1000–100 ps pulse spacing, as discussed above). Although the problem of picosecond cascading of passive nonlinear devices is not yet solved, it is conceivable that their fast switching speeds may be used to make logic decisions (such as gating and wavelength conversion), then amplify the output signals at the same rate, so that the overall bandwidth of the system can still be maintained. The development of high-speed and high-gain laser amplifiers seems promising in providing a possible tool to overcome the pulse energy [2.201]. Thus optical switches may find some applications in fast sequential processing.

Acknowledgements. The results reported in this chapter have been obtained by collaboration with many researchers including H. Gibbs, M. Lindberg, D. Hulin, A. Migus, A. Antonetti, A. Mysyrowicz, H. Haug, L. Banyai, J. Sokoloff, J. Potts, H. Cheng, P. Fluegel, S.H. Park, R. Morgan, M. Joffre, A. Chavez-Pirson, Y. Lee, M. Warren, D. Richardson, R. Jin, V. Williams, F. Jarka, C. Hansen, A. Gossard, H. Morkoç, J. English, D. Sarid and P. Bhattacharya. The funding for this work was obtained from NSF, JSOP, DARPA/RADC, ONR/SDI, NATO, OCC, AFOSR, ARO and grants for computing time were obtained from JVNCC. S.W. Koch is also with the Physics department of the University of Arizona.

References

2.1   C. Klingshirn, H. Haug: Phys. Rep. **70**, 315 (1981)
2.2   A. Miller, D.A.B. Miller, S.D. Smith: Adv. Phys. **30**, 697 (1981)
2.3   B. Hönerlage, R. Levy, J.B. Grun, C. Klingshirn, B. Bohnert: Phys. Rep. **124**, 161 (1985)
2.4   H.M. Gibbs: *Optical Bistability – Controlling Light with Light* (Academic, New York 1985)
2.5   N. Peyghambarian, H.M. Gibbs: Opt. Eng. **24**, 68 (1985)
2.6   S.W. Koch, N. Peyghambarian, H.M. Gibbs: J. Appl. Phys. (Reviews) **63**, R1 (1988)
2.7   S.W. Koch, N. Peyghambarian, M. Lindberg: J. Phys. C (Reviews) **21**, 5229 (1988)
2.8   S. Schmitt-Rink, D.S. Chemla, D.A.B. Miller: Adv. Phys. **38**, 89 (1989)
2.9   H. Haken: *Quantum Field Theory of Solids* (North-Holland, Amsterdam 1976)
2.10   S. Nakajima, Y. Toyozawa, R. Abe: *The Physics of Elementary Excitations*, Springer Ser. Solid-State Sci., Vol. 12 (Springer, Berlin, Heidelberg 1980)
2.11   N.S. Peyghambarian, H. Park, S.W. Koch, A.S. Jeffrey, J.E. Potts, H. Cheng: Appl. Phys. Lett. **52**, 182 (1988)
2.12   Y.H. Lee, A. Chavez-Pirson, S.W. Koch, H.M. Gibbs, S.H. Park, J. Morhange, A. Jeffrey, N. Peyghambarian, L. Banyai, A.C. Gossard, W. Wiegmann: Phys. Rev. Lett. **57**, 2446 (1986)
2.13   O. Madelung, M. Schulz, H. Weiss (eds): Landolt-Börnstein, Vols. III/17 and III/22, Semiconductors (Springer, Berlin, Heidelberg 1982–1987)
2.14   P.W. Smith: Opt. Eng. **19**, 456 (1980)
2.15   P.W. Smith, W.J. Tomlinson: IEEE Spectrum **18**, 16 (1981)
2.16   Y. Toyozawa: Solid State Commun. **28**, 533 (1978)
2.17   E.M. Epshtein: Sov. Phys. – Tech. Phys. **23**, 983 (1978)
2.18   K. Bohnert, H. Kalt, C. Klingshirn: Appl. Phys. Lett. **43**, 1088)1983)
2.19   H. Rossmann, F. Henneberger, I. Voigt: Phys. Status Solidi B **115**, K63 (1983)
2.20   H.X. Nguyen, R. Zimmermann: Phys. Status Solidi B **124**, 191 (1984)
2.21   D.A.B. Miller: J. Opt. Soc. Am B **1**, 857 (1984)
2.22   H.E. Schmidt, H. Haug, S.W. Koch: Appl. Phys. Lett. **44**, 787 (1984)
2.23   S.W. Koch: Lecture Notes in Physics, Vol. 207 (Springer, Berlin, Heidelberg 1984)
2.24   J.A. Goldstone, E. Garmire: Phys. Rev. Lett. **53**, 910 (1984)
2.25   J.W. Haus, C.C. Sung, C.M. Bowden, J.M. Cook: In Proc. of OM 85, *Basic Properties of Optical Material*, ed. by A. Feldman (US Dept. of Commerce, National Bureau of Standards, 1985)
2.26   H.M. Gibbs, G.R. Olbright, N. Peyghambarian, H.E. Schmidt, S.W. Koch, H. Haug: Phys. Rev. A **32**, 692 (1985)
2.27   H. Haug, S.W. Koch, M. Lindberg: Phys. Scr. T **13**, 178 (1986)
2.28   S.W. Koch, H.E. Schmidt, H. Haug: J. Lumin. **30**, 232 (1985)
2.29   A.I. Ekimov, A.L. Efros, A.A. Onushchenko: Solid State Commun. **56**, 921 (1985)
2.30   A.V. Alivisatos, A.L. Harris, N.J. Levinos, M.L. Steigerwald, L.E. Brus: J. Chem. Phys. **89**, 4001 (1988)
2.31   N.F. Borrelli, D.W. Hall, H.J. Holland, D.W. Smith: J. Appl. Phys. **61**, 5399 (1987)
2.32   N. Peyghambarian, S.W. Koch: Rev. Phys. Appl. **22**, 1711 (1987)
2.33   C. Flytzanis, F. Hache, D. Richard, P. Roussignol: Digest of the Intl. Quantum Electron. Conf. (Optical Society of America, Washington, DC 1986) Paper TUNN 1
2.34   K. Kash, A. Scherer, J.M. Worlock, H.G. Craighead, M.C. Tamargo: Appl. Phys. Lett. **49**, 1043 (1986)
2.35   M.A. Reed, R.T. Bate, K. Bradshaw, W.M. Duncan, W.R. Frensley, J.W. Lee, H.D. Shih: J. Vac. Sci. Technol. **4**, 358 (1986)
2.36   J. Cibert, P.M. Petroff, G.J. Dolan, S.J. Pearton, A.C. Gossard, J.H. English: Appl. Phys. Lett. **49**, 1275 (1986)
2.37   Al. L. Efros, A.L. Efros: Sov. Phys. – Semicond. **16**, 772 (1982)
2.38   L.E. Brus: J. Chem. Phys. **80**, 4403 (1984)
2.39   L. Banyai, I. Galbraith, C. Ell, H. Haug: Phys. Rev. B **36**, 6099 (1987)
2.40   E. Hanamura: Solid State Commun. **62**, 465 (1987)
2.41   S. Schmitt-Rink, D.S. Chemla, D.A.B. Miller: Phys. Rev. B **32**, 6601 (1985)
2.42   D.S. Chemla, D.A.B. Miller: Opt. Lett. **11**, 522 (1986)
2.43   Y.Z. Hu, S.W. Koch, M. Lindberg, N. Peyghambarian: In *Nonlinear Optical Materials and Devices for Photonic Switching*, Proc. SPIE **1216**, (1990); Phys. Rev. B (to be published)

2.44   M. Rösler, R. Zimmermann: Phys. Status Solidi B **83**, 85 (1977)
2.45   H. Haug, S. Schmitt-Rink: J. Opt. Soc. Am. B **2**, 1135 (1985)
2.46   M. Lindberg, S.W. Koch: Phys. Rev. B **38**, 3342 (1988)
2.47   D. Sarid, N. Peyghambarian, H.M. Gibbs: Phys. Rev. B., Rapid Commun. **28**, 1184 (1983)
2.48   R. Zimmermann, K. Kilimann, W.D. Kraeft, D. Kremp, G. Röpke: Phys. Status Solidi B **90**, 175 (1978)
2.49   R. Zimmermann: In Conf. Proc., Intl. Conf. on Excitons 84, Guestrew, GDR (1984) (in German)
2.50   K. Kilimann: In Conf. Proc., Intl. Conf. on Excitons 84, Guestrew, GDR (1984) (in German)
2.51   H. Haug, S. Schmitt-Rink: Prog. Quantum Electron. **9**, 3 (1984)
2.52   H. Haug: In *Optical Nonlinearities and Instabilities in Semiconductors*, ed. by H. Haug (Academic, New York 1988)
2.53   L. Banyai, S.W. Koch: Z. Physik B **63**, 283 (1986)
2.54   R.J. Elliott: Phys. Rev. **108**, 1384 (1957)
2.55   J.D. Dow: In Proc. XIIth Conf. Semiconductors, Stuttgart (1974) p. 957
2.56   A.G. Sitenko: In *Lectures in Scattering Theory*, ed. by D. ter Haar, International Series on Monographs in Natural Philosophy (Pergamon, New York 1971)
2.57   S.W. Koch, Y.H. Lee, H.M. Gibbs, N. Peyghambarian: Opt. News **12**, 12 (1986)
2.58   J.F. Morhange, S.H. Park, N. Peyghambarian, A. Jeffrey, H.M. Gibbs, Y.H. Lee, A. Chavez-Pirson, S.W. Koch, A.C. Gossard, J.H. English, M. Masselink, H. Morkoç: Annual Meeting of the Optical Society of America, Seattle, WA (1986)
2.59   Y.H. Lee, A. Chavez-Pirson, B.K. Rhee, H.M. Gibbs, A.C. Gossard, W. Wiegmann: Appl. Phys. Lett. **49**, 1505 (1986)
2.60   M. Wegener, C. Klingshirn, S.W. Koch, L. Banyai: Semicond. Sci. Technol. **1**, 366 (1986)
2.61   M. Dagenais: Appl. Phys. Lett. **43**, 472 (1983)
2.62   D. Hulin, A. Antonetti, L.L. Chase, J.L. Martin, A. Migus, A. Mysyrowicz: Opt. Commun. **42**, 260 (1982)
2.63   A. Antonetti, D. Hulin, A. Migus, A. Mysyrowicz, L.L. Chase: J. Opt. Soc. Am. B **2**, 1197 (1985)
2.64   A. Mysyrowicz, J.B. Grun, R. Levy, A. Bivas, S. Nikitine: Phys. Lett. **26a**, 615 (1968)
2.65   E. Hanamura: Solid State Commun. **12**, 951 (1973)
2.66   G.M. Gale, A. Mysyrowicz: Phys. Rev. Lett. **32**, 727 (1974)
2.67   R. Svorec, L.L. Chase: Solid State Commun. **20**, 353 (1976)
2.68   N. Nagasawa, T. Mita, M. Ueta: J. Phys. Soc. Jpn. **41**, 929 (1976)
2.69   R. Levy, C. Klingshirn, E. Ostertag, V. Duy Phach, J.B. Grun: Phys. Status Solidi B **77**, 381 (1976)
2.70   L.L. Chase, N. Peyghambarian, G. Grynberg, A. Mysyrowicz: Opt. Commun. **28**, 189 (1979)
2.71   A. Maruani, J.L. Oudar, E. Batifol, D.S. Chemla: Phys. Rev. Lett. **41**, 1372 (1978)
2.72   Y. Masumoto, S. Shionoya: Solid State Commun. **38**, 865 (1981)
2.73   L.L. Chase, M.L. Claude, D. Hulin, A. Mysyrowicz: Phys. Rev. A **28**, 3696 (1983)
       D.S. Chemla, D.A.B. Miller: J. Opt. Soc. Am. B **2**, 1155 (1985)
2.74   T. Itoh, T. Kathono: J. Phys. Soc. Jpn. **51**, 707 (1982)
2.75   G.R. Olbright, N. Peyghambarian: Solid State Commun. **58**, 337 (1986)
2.76   N. Peyghambarian, G.R. Olbright, B. Fluegel, S.W. Koch: Digest of the Intl. Quantum Electron. Conf. (Optical Society of America, Washington, DC 1986) postdeadline paper PD20
       N. Peyghambarian, G.R. Olbright, D.A. Weinberger, H.M. Gibbs, B.D. Fluegel: J. Lumin. **35**, 241 (1986)
2.77   R. März, S. Schmitt-Rink, H. Haug: Z. Phys. B **40**, 9 (1980)
2.78   H. Haug, R. März, S. Schmitt-Rink: Phys. Lett. **77A**, 287 (1980)
2.79   S.W. Koch, H. Haug: Phys. Rev. Lett. **46**, 450 (1981)
2.80   E. Hanamura: Solid State Commun. **38**, 939 (1981)
       E. Hanamura: In *Optical Bistability IV*, ed. by N. Peyghambarian, W. Firth, A. Tallet (Editions de Physique, Paris 1988)
2.81   R. Levy, J.Y. Bigot, B. Hönerlage, F. Tomasini, J.B. Grun: Solid State Commun. **48**, 705 (1983)
2.82   N. Peyghambarian, H.M. Gibbs, M.C. Rushford, D. Weinberger: Phys. Rev. Lett. **51**, 1692 (1983)
2.83   C.C. Sung, C.M. Bowden: Phys. Rev. A **29**, 1957 (1984)
2.84   M. Kuwata, T. Mita, N. Nagasawa: Opt. Commun. **40**, 208 (1982)
       M. Lax, W.H. Louisell, W.N. McKnight: Phys. Rev. A **11**, 1365 (1975)

2.85   D.A.B. Miller: IEEE J. QE-17, 306 (1981)
2.86   C.D. Poole, E. Garmire: Appl. Phys. Lett. 44, 363 (1984)
2.87   A. Miller: In Proc. for Meeting on Optical Bistability, Dynamical Nonlinearity and Photonic Logic (Royal Society, London 1984)
2.88   G. Olbright, N. Peyghambarian, H.M. Gibbs, A. Macleod, F. Van Milligen: Appl. Phys. Lett. 45, 1031 (1984)
2.89   S.D. Smith, J.G.H. Mathew, M.R. Taghizadeh, A.C. Walker, B.S. Wherrett, A. Hendry: Opt. Commun. 51, 357 (1984)
2.90   D.A.B. Miller, D.S. Chemla, T.C. Damen, A.C. Gossard, W. Wiegmann, T.H. Wood, C.A. Burrus: Appl. Phys. Lett. 45, 13 (1984)
2.91   M. Dagenais, W.F. Sharfin: Appl. Phys. Lett. 45, 210 (1984)
2.92   M. Shinada, S. Sugano: J. Phys. Soc. Jpn. 21, 1936 (1966)
2.93   R. Dingle, W. Wiegmann, C.H. Henry: Phys. Rev. Lett. 33, 827 (1974)
2.94   J.E. Zucker, A. Pinczuk, D.S. Chemla, A.C. Gossard, W. Wiegmann: Phys. Rev. Lett. 51, 1293 (1983)
2.95   D.S. Chemla, D.A.B. Miller, S. Schmitt-Rink: In Optical Nonlinearities and Instabilities in Semiconductors, ed. by H. Haug (Academic, Cambridge 1988) p. 83
2.96   H. Stolz: Einführung in die Vielelektronentheorie der Kristalle (Akademie Verlag, Berlin 1974)
2.97   D.A.B. Miller, D.S. Chemla, D.J. Eilenberger, P.W. Smith, A.C. Gossard, W. Wiegmann: Appl. Phys. Lett. 42, 925 (1983)
2.98   M.W. Derstine, D.E. Grider, J.A. Lehman, P.P. Ruden, N. Peyghambarian: Proc. SPIE Conf. on Optical Computing and Nonlinear Materials, ed. by N. Peyghambarian, Vol. 881, (SPIE, Los Angeles, CA 1988) p. 131
2.99   H.C. Lee, A. Hariz, P.D. Kapkus, A. Kost, M. Kawase, E. Garmire: Appl. Phys. Lett. 50, 1182 (1987)
2.100 S.H. Park, J.F. Morhange, A.D. Jeffery, R.A. Morgan, A. Chavez-Pirson, H.M. Gibbs, S.W. Koch, N. Peyghambarian, M. Derstine, A.C. Gossard, J.H. English, W. Wiegmann: Appl. Phys. Lett. 52, 1201 (1988)
2.101 G.H. Döhler, H. Kunzel, D. Olego, K. Ploog, P. Ruden, H. Stolz, G. Abstreiter: Phys. Rev. Lett. 47, 864 (1981)
2.102 P. Ruden, G.H. Döhler: Phys. Rev. B 27, 2538 (1983)
2.103 G. Döhler: IEEE J QE-22, 1682 (1986)
2.104 P. Ruden: Proc. of SPIE Conf. on Quantum Well and Superlattice Physics, ed. by G.H. Döhler, J.N. Schulman, SPIE, Vol. 792 (1987) p. 36
2.105 J.S. Yuan, M. Gal, P.C. Taylor, G.B. Springfellow: Appl. Phys. Lett. 47, 405 (1985)
2.106 K.W. Carey, G.H. Dohler, J. Turner, J. Vilms: Proc. of the 12th Intl. Symp. on GaAs and Related Compounds, 1985. Inst. of Phys. Conf. Ser. 79, 385 (1986)
2.107 K. Nakagawa, Y. Shiraki: Surf. Sci. 174, 646 (1986)
2.108 P. Jantsch, G. Bauer, P. Pichler, H. Clemens: Appl. Phys. Lett. 47, 738 (1985)
2.109 A. Kost, E. Garmire, A. Danner, P.D. Dapkus: Appl. Phys. Lett. 52, 837 (1988)
2.110 L.A. Kolodziejski, R.L. Gunshor, N. Otsuka, S. Datta, W.M. Becker, A.V. Nurmikko: IEEE J QE-22, 1666 (1986)
2.111 D.R. Anderson, R.L. Gunshor, L.A. Kolodziejski, S. Datta, A.E. Kaplan, A.V. Nurmikko: Proc. 1985 Mater. Res. Soc. Meeting, Boston, MA (1985) p. 81
2.112 P.M. Petroff, A.C. Gossard, R.A. Logan, W.W. Wiegmann: Appl. Phys. Lett. 41, 635 (1982)
2.113 A.B. Fowler, A. Hartstein, R.A. Webb: Phys. Rev. Lett. 48, 196 (1982)
2.114 B.I. Green, J. Orensetin, R.R. Millard: Phys. Rev. Lett. 58, 2750 (1987)
2.115 G.W. Bryant: Phys. Rev. B 29, 6632 (1984)
2.116 M.J. Tsuchiya, M. Gaines, R.H. Yan, R.J. Simes, P.O. Holtz, L.A. Coldren, P.M. Petroff: Phys. Rev. Lett. 62, 466 (1989)
2.117 M. Tanaka, H. Sakaki: Appl. Phys. Lett. 54, 1326 (1989)
2.118 A. Forchel, G. Tränkle, U. Cebulla, H. Leier, B.E. Maile: In Optical Switching in Low-Dimensional Systems, ed. by H. Haug, L. Banyai (Plenum, New York 1989) p. 361
2.119 E. Kapon, S. Simhony, D.M. Hwang, K. Kash, R. Bhat, E. Colas: QELS 89 Conference, Baltimore, MD (1989) postdeadline paper
2.120 E. Kapon, M.C. Tamargo, D.M. Hwang: Appl. Phys. Lett. 50, 347 (1987)
2.121 R. Loudon: Am. J. Phys. 44, 1064 (1976)
2.122 L. Brus: IEEE J. QE-22, 1909 (1986) and the references to the author's earlier work
2.123 R.K. Jain, R.C. Lind: J. Opt. Soc. Am. 73, (1983)
2.124 S.S. Yao, C. Karaguleff, A. Gabel, R. Fortenbery, C.T. Seaton, G. Stegeman: Appl. Phys. Lett. 46, 801 (1985)

2.125 P. Roussignol, D. Ricard, K.C. Rustagi, C. Flytzanis: Opt. Commun. **55**, 1431 (1985)
2.126 M.C. Nuss, W. Zinth, W. Kaiser: Appl. Phys. Lett. **49**, 1717 (1987)
2.127 G.R. Olbright, N. Peyghambarian, S.W. Koch, L. Banyai: Opt. Lett. **12**, 413 (1987)
2.128 I.M. Lifshitz, V.V. Slyozov: J. Phys. Chem. Sol. **19**, 35 (1961)
2.129 N. Peyghambarian, S.H. Park, R.A. Morgan, B. Fluegel, Y.Z. Hu, M. Lindberg, S.W. Koch, D. Hulin, A. Migus, J. Etchepare, M. Joffre, G. Grillon, D.W. Hall, N.F. Borrelli: In *Optical Switching in Low-Dimensional Systems*, ed. by H. Haug, L. Banyai (Academic, Orlandeo, FL 1989) p. 191
2.130 N. Peyghambarian, B. Fluegel, D. Hulin, A. Migus, M. Joffre, A. Antonetti, S.W. Koch, M. Lindberg: IEEE J Quant. Elect. Dec. 1989
2.131 E. Hilinski, P. Lucas, Y. Wang: J. Chem. Phys. **89**, 3435 (1988)
2.132 P. Roussignol, D. Ricard, C. Flytzanis, N. Neuroth: Phys. Rev. Lett. **62**, 312 (1989)
2.133 L. Wang, V. Esch, R. Feinleib, L. Zhang, R. Jin, J.M. Chou, R.W. Sprague, H.A. Macleod, G. Khitrova, H.M. Gibbs, K. Wagner, D. Psaltis: Appl. Opt. **27**, 1715 (1988)
2.134 S.H. Park, R.A. Morgan, S.W. Koch, N. Peyghambarian: J. Opt. Soc. Am. B, to be published
2.135 R.A. Morgan, S.H. Park, S.W. Koch, N. Peyghambarian: Semicond. Sci. Technol., Special Issue on Quantum Wells and Optoelectronis (1989)
2.136 P. Roussignol, D. Ricard, J. Lukasik, C. Flytzanis: J. Opt. Soc. Am. B **4**, 5 (1987)
2.137 M. Mitsunaga, H. Shinojima, K. Kubodera: J. Opt. Soc. Am. B **5**, 1448 (1988)
2.138 D. Hall, N.F. Borrelli: J. Opt. Soc. Am. B **5**, 1650 (1988)
2.139 L. Banyai, S.W. Koch: Phys. Rev. Lett. **57**, 2722 (1986)
2.140 T. Takagahara: Proc. of IQEC '88, Tokyo, Japan (1988) paper ThB4, p. 620
2.141 S. Schmitt-Rink, D.A.B. Miller, D.S. Chemla: Phys. Rev. B **35**, 8113 (1987)
2.142 L. Banyai, M. Lindberg, S.W. Koch: Opt. Lett. **13**, 212 (1988)
       N. Bloembergen: *Nonlinear Optics* (W.A. Benjamin, Reading, MA 1965)
2.143 U. Woggon, F. Henneberger: In *Optical Bistability IV*, ed. by W. Firth, N. Peyghambarian, A. Tallet (Edition de Physique, Paris 1988)
2.144 D. Cotter: In *Ultrafast Phenomena V*, ed. by G.R. Fleming, A.E. Siegman, Springer Ser. Chem. Phys., Vol. 46 (Springer-Verlag, Berlin, Heidelberg 1986)
2.145 J. Etchepare, G. Grillon, I. Thomazeau, G. Hamoniaux, A. Orszag: In *Ultrafast Phenomena V*, ed by G.R. Fleming, A.E. Siegman, Springer Ser. Chem. Phys., Vol. 46 (Springer-Verlag, Berlin, Heidelberg 1986)
2.146 N. Peyghambarian, G.R. Olbright, S.W. Koch, B. Fluegel: Proc. of the NSF Workshop on Optical Nonlinearities, Fast Phenomena and Signal Processing, ed. by N. Peyghambarian, Tucson, AZ (1986)
2.147 K. Shum, G.C. Tang, M.R. Junnarkar, R.R. Alfano: Appl. Phys. Lett. **51**, 1839 (1987)
2.148 S.C. Hsu, H.S. Kwok: Appl. Phys. Lett. **50**, 1782 (1987)
2.149 D.G. Steel, J.T. Remillard, H. Wang, M.D. Webb: In Digest of the Conference on Lasers and Electro Optics (Optical Society of America, Washington, DC 1988) paper WQ1
2.150 C.T. Seaton, G.I. Stegeman: Appl. Phys. Lett. **49**, 1403 (1986)
2.151 D. Cotter: Digest of the Intl. Quant. Elect. Conf. (Optical Society of America, Washington, DC 1986) post deadline paper PD 19
2.152 J. Warnock, D.D. Awschalom: Phys. Rev. B **32**, 5529 (1985)
2.153 D. Sarid, W. Gibbons, M. Warren: Digest of the Topical Meeting on Optical Bistability 3 (Optical Society of America, Washington, DC 1985) paper WA5
2.154 N. Finlayson, W.C. Banyai, A. Gabel, K.W. Delong, C.T. Seaton, G.I. Stegeman, J. Bell, T.C. Cullen, C.N. Ironside: Proc. of SPIE Conf. on Optical Computing and Nonlinear Materials, ed. by N. Peyghambarian, Vol. 881 (SPIE, Los Angeles, CA 1988) p. 155
2.155 G. Assanto: In *Nonlinear Optics and Optical Computing*, ed. by A.N. Chester, S. Martellucci (Academic, Cambridge 1988)
2.156 S. Wabnitz: In *Nonlinear Optics and Optical Computing*, ed. by A.N. Chester, S. Martellucci (Academic, Cambridge 1988)
2.157 L. Genzel, T.P. Martin: Surf. Sci. **34**, 33 (1973)
2.158 M. Born, E. Wolf: *Principles of Optics* (Pergamon, New York 1970)
2.159 S.L. McCall: Phys. Rev. A **9**, 1515 (1974)
2.160 R. Bonifacio, L.A. Lugiato: Opt. Commun. **19**, 172 (1976)
2.161 R. Bonifacio, L.A. Lugiato: Phys. Rev. A **18**, 1129 (1978)
2.162 J.H. Marburger, F.S. Felber: Phys. Rev. A **17**, 335 (1978)
2.163 G.P. Agrawal, H.J. Charmichael: Phys. Rev. A **19**, 2074 (1979)

2.164 S.W. Koch: In *Optical Nonlinearities and Instabilities in Semiconductors*, ed. by H. Haug (Academic, New York 1988)

2.165 M. Warren, S.W. Koch, H.M. Gibbs: In IEEE Computer Science Publication "Computer", ed. by T.E. Bacjman, E.A. Parrish (December 1987)

2.166 H.M. Gibbs, S.L. McCall, T.N.C. Venkatesan, A.C. Gossard, A. Passner, W. Wiegmann: Appl. Phys. Lett. 35, 451 (1979)

2.167 H.M. Gibbs, S.S. Tarng, J.L. Jewell, D.A. Weinberger, K. Tai, A.C. Gossard, S.L. McCall, A. Passner, W. Wiegmann: Appl. Phys. Lett. 41, 221 (1982)

2.168 S.S. Tarng, H.M. Gibbs, J.L. Jewell, N. Peyghambarian, A.C. Gossard, T. Venkatesan, W. Wiegmann: Appl. Phys. Lett. 44, 360 (1984)

2.169 H. Kogelnik: In *Integrated Optics*, ed. by T. Tamir, 2nd ed. Topics Appl. Phys., Vol. 7, (Springer, Berlin, Heidelberg 1985)

2.170 V. Ramaswamy: Bell Syst. Tech. J. 53, 697 (1974)

2.171 A. Yariv, P. Yeh: *Optical Waves in Couplets* (Wiley, New York 1984)

2.172 G.I. Stegeman, R.H. Stolen (ed.): *Nonlinear Guided Wave Phenomena*, special issue of J. Opt. Soc. Am. B 5, 264–574 (1988) and references therein

2.173 M. Warren, W. Gibbons, K. Komatsu, D. Sarid, D. Hendricks, H.M. Gibbs, J.M. Sugimoto: IQEC '87, Baltimore, MD, April 1987, Postdeadline paper PD9

2.174 P. LiKamwa, J.E. Sitch, N.J. Mason, J.S. Roberts, P.N. Robson: Electron. Lett. 22, 1129 (1986)

2.175 U. Das, Y. Chen, B. Bhttacharya: Appl. Phys. Lett. 51, 1679 (1987)

2.176 R. Jin, C.C. Chuang, H.M. Gibbs, S.W. Koch: Appl. Phys. Lett. 53, 1791 (1988)

2.177 W.M. Gibbons, D. Sarid: Appl. Phys. Lett. 51, 403 (1987)

2.178 R.L. Jin, R. Wang, W. Sprague, H.M. Gibbs, G. Gigioli, H. Kulcke, H.A. Macleod, N. Peyghambarian, G.R. Olbright, M. Warren: In *Optical Bistability III*, ed. by H.M. Gibbs, P. Mandel, N. Peyghambarian, S.D. Smith, Springer Proc. Phys., Vol. 8 (Springer, Berlin, Heidelberg 1986)

2.179 F.V. Karpushko, G.V. Sinitsyn: Appl. Phys. B 28, 137 (1982)

2.180 S.D. Smith, I. Janossy, H.A. MacKenzie, J.G.H. Mathew, J.J.E. Reid, M.R. Taghizadeh, F.A.P. Tooley, A.C. Walker: Opt. Eng. 24, 569 (1985)

2.181 S.D. Smith, A.C. Walker, F.A.P. Tooley, J.H. Mathew, M.R. Taghizadeh: In *Optical Bistability III*, ed. by H.M. Gibbs, P. Mandel, N. Peyghambarian, S.D. Smith, Springer Proc. Phys., Vol. 8 (Springer, Berlin, Heidelberg 1985)

2.182 H.M. Gibbs, N. Peyghambarian: *Nonlinear Etalons and Optical Computing*, International Commission on Optics, Optical Computing Conf., Jerusalem (1986)

2.183 M.T. Tsao, L. Wang, R. Jin, R.W. Sprague, G. Gigioli, H.M. Kulcke, Y.D. Li, H.M. Chou, H.M. Gibbs, N. Peyghambarian: Opt. Eng. 26, 41 (1987)

2.184 J.L. Jewell, Y.H. Lee, M. Warren, H.M. Gibbs, N. Peyghambarian, A.C. Gossard, W. Wiegmann: Appl. Phys. Lett. 46, 918 (1985)

2.185 T. Venkatesan, B. Wilkens, M. Warren, Y.H. Lee, G. Olbright, H.M. Gibbs, N. Peyghambarian, J.S. Smith, A. Yariv: Appl. Phys. Lett. 48, 145 (1986)

2.186 J.L. Jewell, A. Scherer, S.L. McCall, A.C. Gossard, J.H. English: Appl. Phys. Lett. 51, 94 (1987)

2.187 A. Migus, A. Antonetti, D. Hulin, A. Mysyrowicz, H.M. Gibbs, N. Peyghambarian, J.L. Jewell: Appl. Phys. Lett. 46, 70 (1985)

2.188 Y.H. Lee, M. Warren, G. Olbright, H.M. Gibbs, N. Peyghambarian, T. Venkatesan, J.S. Smith, A. Yariv: Appl. Phys. Lett. 48, 754 (1986)

2.189 J.L. Jewell, Y.H. Lee, J.F. Duffy, A.C. Gossard, W. Wiegmann, J.H. English: In *Optical Bistability III*, ed. by H.M. Gibbs, P. Mandel, N. Peyghambarian, S.D. Smith, Springer Proc. Phys., Vol. 8 (Springer, Berlin, Heidelberg 1986)

2.190 M. Ojima, A. Chavez-Pirson, Y.H. Lee, J.F. Morhange, H.M. Gibbs, N. Peyghambarian, F.-Y. Juang, P.K. Bhattacharya, D.A. Weinberger: Appl. Opt. 25, 2311 (1986)

2.191 D. Hulin, A. Mysyrowicz, A. Antonetti, A. Migus, W.T. Masselink, H. Morkoç, H.M. Gibbs, N. Peyghambarian: Appl. Phys. Lett. 49, 749 (1986)

2.192 J.L. Jewell, A. Scherer, S.L. McCall, A.C. Gossard, J.H. English: PDP1-1, Topical Meeting on Photonic Switching, Incline Village, Nevada (1987)

2.193 W.F. Sharfin, M. Dagenais: Appl. Phys. Lett. 46, 819 (1985)

2.194 F.A.P. Tooley, S.D. Smith, C.T. Seaton: Appl. Phys. Lett. 43, 807 (1983)

2.195 D.A.B. Miller, S.D. Smith, A. Johnston: Appl. Phys. Lett. 35, 658 (1979)

2.196 B.K. Jenkins, A.A. Sawchuk, T.C. Strand, R. Forchheimer, B.H. Soffer: Appl. Opt. **23**, 3455 (1984)
2.197 R. Jin, C. Hansen, M. Warren, D. Richardson, H.M. Gibbs, N. Peyghambarian, G. Khitrova, S.W. Koch: Appl. Phys. B **46**, 61 (1988)
2.198 L. Wang, W.Z. Geng, R. Kostuk: Holographic Optics: Optically and Computer Generated. Proc. SPIE **1052**, 97 (1988)
2.199 A. Huang: Proc. IEEE 10th Intl. Optical Computing Conf., Publication No. 83, CH1880-4 (1983) pp. 13–17
2.200 L. Zhang, R. Jin, C.W. Stirk, G. Khitrova, R.A. Athale, H.M. Gibbs, H.M. Chou, R.W. Sprague, H.A. Macleod: To be published
2.201 J. Hegarty, K.A. Jackson: Appl. Phys. Lett. **45**, 1314 (1984)

3. Optical Interconnects

R.K. Kostuk, J.W. Goodman, and L. Hesselink

With 14 Figures

A computational system relies heavily on the connections of its components. This is true for both general purpose electronic and special purpose optical computers. Although both systems are presently topics of intense research, this chapter will emphasize interconnection techniques for digital computers based on electronic integrated switching circuits. This is done for two reasons. First, electronic computers will, in all probability, be the most commonly used machines for some time to come. Therefore, the successful application of optics in this area could have a much broader technological impact than by its application to a limited number of special purpose optical computers. Second, the switching components for electronic systems are readily available. Therefore, optics would only have to address one problem at the present time. Also, the techniques used in connecting electronic systems could eventually be applied to digital optical computers if nonlinear optical switching elements become competitive with their electronic counterparts.

Interconnection is primarily a communications issue. Information must be transferred to various levels of the computing systems at different data rates. These levels have been established in part by the evolution of integration and packaging technology. At the highest level of connection are machine-to-machine connections. Optical links at this level are commercially available in the form of fiber optic local area networks. Data is transferred in serial fashion over a common optical bus. The next level of connection is at the board-to-board level where some preliminary experimental work has been demonstrated [3.1]. Further down on the scale are connections at the chip-to-chip level, and finally at the device-to-device or intra-chip levels.

Interconnects may also be categorized by the type of communication link or the function of the connection. Several important types are [3.2]: 1) internal point-to-point links required for intra-chip connections between logic and memory; 2) external point-to-point connections for applications such as loading information from an external device to memory; 3) internal global connections for broadcasting an internal signal to all other functions in the system; and 4) external-to-internal global communications needed for distributing an external clock signal to different locations on a chip. The terms internal and external are relative to a specific level of integration.

Two factors will determine how far down into the integration hierarchy optics will perform these functions. The first are the significant advantages in performance which optics can offer over electrical interconnects, and the second is the

Springer Series in Electronics and Photonics, Vol. 30   61
Nonlinear Photonics Editors: H.M. Gibbs · G. Khitrova · N. Peyghambarian
© Springer-Verlag Berlin Heidelberg 1990

ability to implement these connections in a reliable and cost-effective manner. This chapter will concentrate on issues related to these two factors. First, the potential advantages of optical interconnects are discussed. Then a comparison of electrical and optical interconnects is presented. This will be used to identify specific areas where optics has an advantage over electrical connections. Next, general methods for distributing optical signals by guided and free-space optical elements are outlined, followed by a detailed analysis of a holographic interconnect design.

3.1 Potential Advantages of Optical Interconnects

A number of factors have motivated the study of optical interconnects as a replacement for electrical connections. In this section, a number of beneficial and limiting features of optical techniques for interconnect applications are discussed.

One of the most important advantages of optics for this application is immunity to mutual interference effects. When electrical lines intersect time varying electromagnetic fields produced by adjacent lines, inductive and capacitive coupling results. This effect couples signals between lines and reduces the noise margin of the received information. These problems increase with higher signal bandwidths and greater interconnect line densities.

In contrast, two beams of light can propagate directly through each other without coupling as long as the propagation medium is linear. This is a consequence of the fact that photons are bosons and can occupy the same cell of phase space, while electrons, which are fermions, cannot. Although optical propagation paths have this advantage, interference effects may still result between electrical logic circuits and optical source drivers. This effect can limit the packing density of the hybrid electro-optical system.

The minimum propagation delay of an electromagnetic signal is determined by the speed of light in the propagation medium. In theory then, optical signals can readily propagate at the maximum realizable speed. Other factors can also affect the transmission of information. For instance, a laser driver required to transmit an optical signal, and an electrical transmission line driver, both have rise-time characteristics which must be considered in determining the speed of a transmitted signal. Therefore, a comparison of the two transmission methods must include all elements of the communication link.

Another potential advantage of optical interconnects is freedom from planar and quasiplanar constraints. Conventional integrated circuit design techniques restrict interconnects to a few layers, and within each layer lines cannot cross. Optical waveguides can cross provided the angle between the guides exceeds about 10°, and free space interconnects can also cross without significant interference. These factors can provide a new degree of flexibility in both the design and the architecture of computational systems.

VLSI chip-to-chip interconnections require several hundred mechanical bonds between signal lines and bonding pads. As the number of transistors per chip increases, the number of mechanical bonds must also increase, which reduces the reliability of the system. A free space optical interconnect system does not require a mechanical contact for each point. This could provide an advantage in the reliability of the system, provided that mounts for the optical elements are stable over the operating range of the computer.

The final possible advantage of optics for interconnections is the potential for using dynamical optical components as reconfigurable interconnects. A system of this type would have significant impact on Fourier transform algorithms, sorting operations, and adaptive processing techniques which are difficult to realize with electrical interconnects. However, although preliminary demonstrations of this concept have been performed with phase conjugating mirrors [3.3], packaging, efficiency, and reliability issues of currently available materials must be circumvented before practical implementation of these systems is possible.

3.2 Comparison of Electrical and Optical Interconnections

Several characteristics can be used to compare the performance of different interconnections [3.2]. These include power requirements, data rate, signal delay, switching energy or power-delay product, and interconnect density/area. It is difficult to specify which of these criteria will have the most significant effect on all interconnect systems. However, a fundamental limit on design is the power/area that can be dissipated by the semiconductor material. Therefore, a comparison of electrical and optical interconnects will be performed on the basis of respective power requirements which are necessary to transmit equivalent signals. These results can then be combined with spatial requirements of the devices to evaluate the power/area constraints.

3.2.1 Electrical Interconnect Power Requirements

A basic power unit at the intra-chip connection level is the power needed by one electronic inverter to trigger another inverter. When these devices are separated by distances that are relatively short compared to the wavelength of the electromagnetic field being transmitted, the power required by the driving circuit is primarily reactive in nature, and is given by

$$P = \frac{CV^2}{2\tau},$$
(3.1)

where C is the capacitance of the line and the attached device, V is the electronic device threshold level, and τ is the clocking period. Figure 3.1 shows a typical gate-to-gate connection [Ref. 3.4, p. 10]. In order to produce a response,

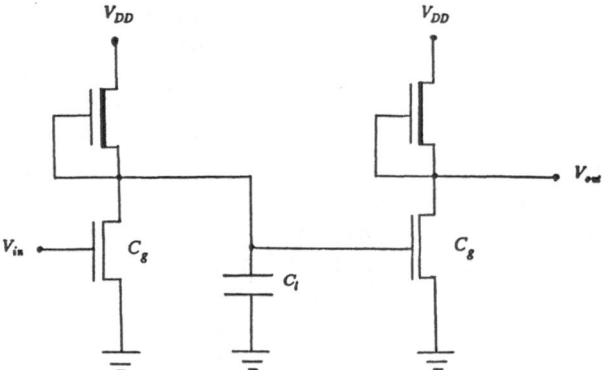

Fig. 3.1. Electrical gate-to-gate connection. C_g: gate capacitance; C_l: line capacitance

the capacitance of the two devices and the line connecting them must be charged to the threshold potential of the gate. The total capacitance of this connection becomes

$$C_{tg} = 2C_g + C_l , \tag{3.2}$$

where C_{tg} is the total capacitance of the gate-to-gate interconnection, C_g is the gate capacitance, and C_l is the capacitance of the line connecting the two gates.

In terms of the device geometry and dielectric coefficients, this expression becomes

$$C_{tg} = 2 \left[\epsilon_r \epsilon_0 \frac{w_g l_g}{t_{0g}} \right] + \epsilon_r \epsilon_0 \frac{w_l l_l}{t_l} , \tag{3.3}$$

with w the width, l the length, and t the thickness of either the gate (g), or the line (l), ϵ_r is the relative permittivity of the dielectric (for SiO$_2$ $\epsilon_r = 3.9$), and $\epsilon_0 = 8.854 \times 10^{-14}$ F/cm.

The average length of long-distance on-chip interconnections is approximately given by [3.5]

$$l_l = \frac{\sqrt{\text{Chip Area}}}{2} = \frac{\sqrt{A_c}}{2} ; \tag{3.4}$$

therefore,

$$C_{tg} = \epsilon_r \epsilon_0 \left[2 \frac{w_g l_g}{t_{0g}} + \frac{w_l \sqrt{A_c/2}}{t_l} \right] . \tag{3.5}$$

Field fringing effects restrict the minimum line width/height ratio to a value of $\simeq 2$ [3.6], which leads to

$$C_{tg} = \epsilon_r \epsilon_0 \left[2 \frac{w_g l_g}{t_{0g}} + \sqrt{2A_c} \right] . \tag{3.6}$$

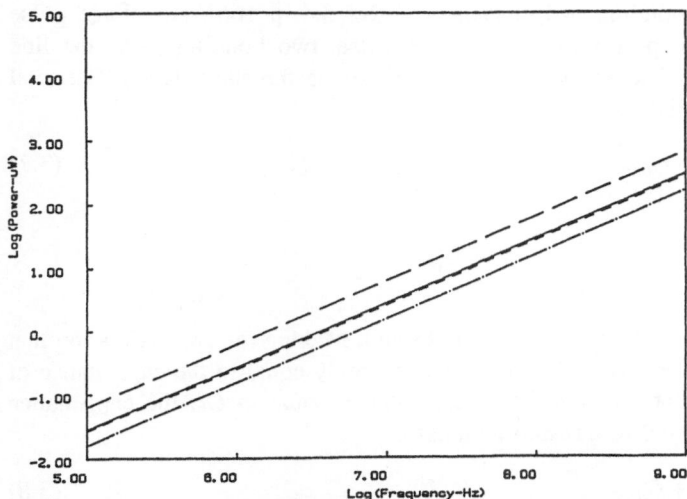

Fig. 3.2. Log (power [μW]) vs log (frequency [Hz]) for gate-to-gate and chip-to-chip interconnections. (—) 1 μm gate, 1.0 cm^2 chip; (– · –) 1 μm gate, 0.5 cm^2 chip; (- - -) 0.5 μm gate, 1.0 cm^2 chip; (– – –) chip-to-chip connection

Typical device dimensions for a 1 μm linewidth silicon MOS transistor [3.6] are l_g = 1.0 μm, w_g = 5.0 μm, and t_{0g} = 0.005 μm. With a total chip area A_c = 1.0 cm^2, the capacitance of the connection becomes C_{1tg} = 560 fF. The power needed for this interconnection as a function of driving frequency is compared to that required by a connection with the same area, but with a 0.5 μm linewidth (C_{2tg}), and a system with the same 1.0 μm linewidth, but a 0.5 cm^2 chip area (C_{3tg}). A log-log plot of the required power vs frequency for each case is shown in Fig. 3.2. At 1 GHz, the required electrical drive power for C_{1tg} is 280 μW; C_{2tg}, 260 μW; and C_{3tg}, 156 μW. This indicates that the required power is most sensitive to the area of the chip, and scales approximately linearly with the chip length.

Figure 3.3 shows a chip-to-chip connection. To minimize propagation delays, the gate capacitance is gradually increased until the device capacitance is

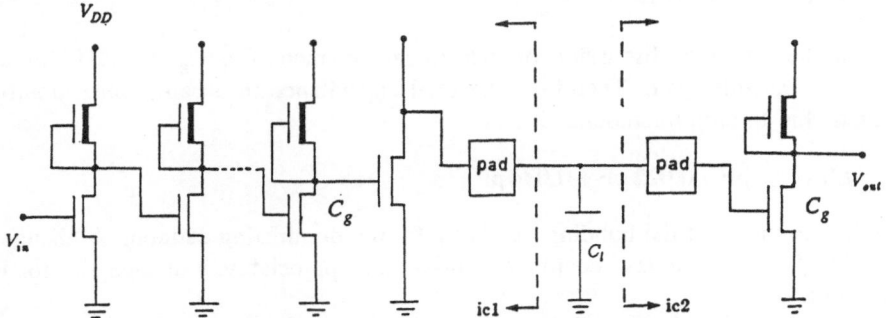

Fig. 3.3. Electrical chip-to-chip connection. C_g: gate capacitance of last transistor in drive circuit; C_l: line capacitance

65

comparable to the bonding pad capacitance [Ref. 3.4, p. 166]. A voltage pulse must have sufficient power to charge these gates, two bonding pads, the line connecting them, and a receiving gate to the device threshold level. The total capacitance of this link is

$$C_{tl} = 2C_b + C_l + C_{driver} + C_g , \qquad (3.7)$$

where

$$C_b = \epsilon_r \epsilon_0 \frac{A_{pad}}{t_{0b}} ,$$

and C_{driver} is the capacitance of the driver circuit. Making the approximation that the capacitance of the driver circuit is approximately equal to the capacitance of the last gate, and that this gate has a capacitance equal to C_b, the capacitance which must be charged to transmit a signal is

$$C_{tl} \simeq 3C_b + C_l + C_g . \qquad (3.8)$$

A typical bonding pad is about $100 \, \mu m$ in length, and assuming an oxide separation from the ground plane of $1 \, \mu m$, $C_b = 0.4 \, pF$. The capacitances of the bonding wires are again related to the area and the height of the lines above the substrate

$$C_l = \epsilon_r \epsilon_0 \frac{w_l l_l}{t_{0l}} .$$

The term t_{0l} is assumed to be equal to the thickness of a silicon wafer, and is on the order of $0.5 \, mm$. The width of bonding wire is about $25 \, \mu m$, which gives a capacitance per unit line length of

$$C_l = 4.4 l_l \, [fF] ,$$

where the line length l_l is in centimeters. Therefore, for line lengths between 1 and 5 cm, the line capacitance has values

$$4.4 \, fF < C_l < 22 \, fF .$$

From the previous discussion of gate-to-gate connections, $C_g \simeq 34 \, fF$ for a $1.0 \, \mu m$ linewidth gate. Therefore, the total capacitance of a capacitance dominated chip-to-chip interconnection is

$$C_{tl} = 1.2 \, pF + 0.022 \, pF + 0.034 \, pF ,$$

which implies that the bonding pad is by far the dominating element. A change in line length by a few centimeters does not appreciably influence the total capacitance.

The corresponding power required to transmit a signal on this line is plotted in Fig. 3.2. At a transmission frequency of 1 GHz, about $625 \, \mu W$ of reactive power is required. If the interconnection is formed by a matched, lossless transmission line, the average power required to transmit a signal is

$$P_{tt} = DF \frac{|V|^2}{Z_0} \,, \tag{3.9}$$

where DF is the duty factor of the transmitted pulse code, and

$$Z_0 = \sqrt{\frac{L_0}{C_0}} \,,$$

is the line impedance with L_0 the characteristic line inductance per unit length, and C_0 the capacitance per unit line length. For high speed digital systems, typical values for Z_0 range from 30 to 200 Ω [3.7]. With a duty factor of 50 %, a threshold voltage of 1 V, and $Z_0 = 50\,\Omega$, the average power dissipated in transmitting a signal is 10 mW. Note that, since the line has been assumed lossless, this required drive power also applies for a terminated superconducting transmission line.

For this ideal case, the required average power is independent of both line length and the frequency of the transmitted signal, implying that all harmonic frequency components of the transmitted pulse are absorbed by the load. In actual situations, impedance mismatching and attenuation effects will increase the required power.

3.2.2 Optical Interconnects

Next, consider a simple electro-optic link consisting of a semiconductor source and detector. For this comparison, it is assumed that all the light from the source is focused on the detector. The detector circuit model is shown in Fig. 3.4. The current generated is a function of the physical parameters of the junction and the illumination [3.8]

$$i_p = \Phi \frac{q(1-r)}{h\nu}(1 - e^{-\alpha_0 z}) = \Phi R_d \,, \tag{3.10}$$

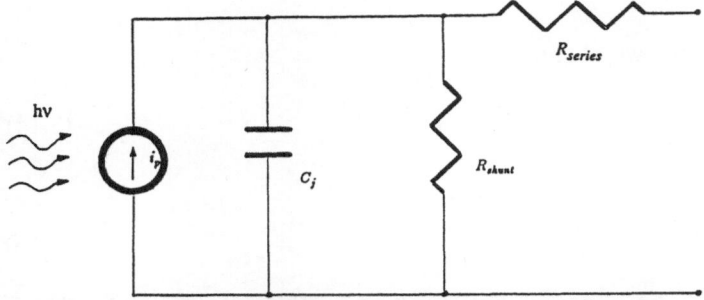

Fig. 3.4. Equivalent circuit for a photodiode. i_p: current source equivalent of photocurrent; C_j: junction capacitance; R_{shunt}: shunt resistance; R_{series}: series resistance

where i_p is the photocurrent, Φ is the optical flux illuminating the active area, q is the charge of an electron, r is the Fresnel reflection coefficient of the detector surface, $h\nu$ is the photon energy, α_0 is the semiconductor absorption coefficient at wavelength λ, z is the absorption depth, and R_d is the detector responsivity. A typical value for the responsivity of a silicon detector at 830 nm is about 0.45 A/W.

A reasonably well fabricated silicon detector has a large shunt and low series resistance, which simplifies the model to a capacitance shunting a current source. Current from the detector must charge the gate to its threshold level in a time less than the shortest period τ of a transmitted pulse code.

If no preamplification is assumed, all current must originate from electrons generated from the incident optical flux, Φ, illuminating the junction of the detector. The depth of this junction t_{det} is limited by the transit time of the generated carriers. A detector with a response time on the order of a nanosecond has a depth of about $2\,\mu m$ [3.9]. This depth and the area of the detector A_{det}, set the junction capacitance of the detector according to the relationship

$$C_d = \epsilon_r \epsilon_0 \frac{A_{det}}{t_{det}} . \tag{3.11}$$

The total capacitance which must be charged to the 1 V threshold is the sum of the detector junction and gate capacitances. (It is assumed that the line connecting the detector and gate is short and introduces a relatively small additional capacitance.) Therefore

$$C_{td} = C_d + C_g . \tag{3.12}$$

In a given pulse period τ, an optical source must provide sufficient flux, Φ, at the detector to generate current and charge the capacitance C_{td} to the threshold voltage. The required mathematical relations are

$$Q = \int_0^\tau i\,dt \simeq i\tau , \tag{3.13}$$

and

$$V_{th} = \frac{Q}{C_{td}} .$$

Using (3.10), thus we obtain

$$V_{th} = \frac{i\tau}{C_{td}} = \frac{R_d \Phi \tau}{C_{td}} , \tag{3.14}$$

or,

$$\Phi = \frac{V_{th} C_{td}}{R_d \tau} . \tag{3.15}$$

Figure 3.5 shows the optical power required to charge the detector and gate to a 1 V threshold as a function of modulation frequency for different detector

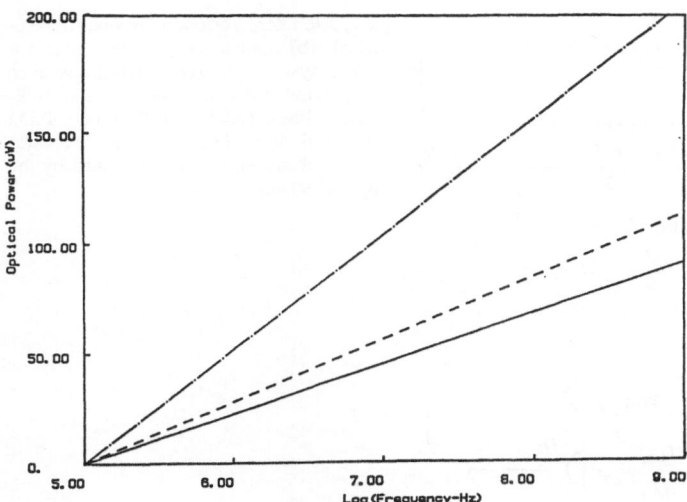

Fig. 3.5. Required optical power [μW] vs log (frequency [Hz]) to produce 1.0 V threshold voltage from detectors with (—) 10 μm; (- - - -) 25 μm; and (----) 100 μm detector diameters

areas. These areas correspond to detector lengths of 10, 25, and 100 μm. As shown, a 100 μm detector requires about 200 μW of optical power to charge the link at a gigahertz rate, and as the detector area is reduced to about 10 μm, the required power approaches that necessary to charge the gate capacitance alone (about 86 μW).

3.2.3 Fan-Out

The above analysis can also be extended to the problem of addressing several receiving locations simultaneously. This is the problem of a signal fan-out, and is an important consideration for parallel processing and clock distribution applications. The situation is illustrated in Fig. 3.6a. When lines are short compared to the transmitted wavelength, the capacitance of the unterminated branched connections can be expressed by

$$C_t = (C_b + C_g) + M(C_l + C_g + C_b) \, , \tag{3.16}$$

where the first term in parentheses represents the gate and bonding pad capacitance of the driver, and M is the number of lines being driven. When the line lengths are comparable with the transmission wavelength, the ideal terminated line power required is

$$P = \frac{V^2}{2} \frac{1}{Z_0'} = \frac{V^2}{2} \frac{M}{Z_0} \, , \tag{3.17}$$

where Z_0 is the impedance of an individual line. In each case, the increased power is proportional to the number of lines being driven.

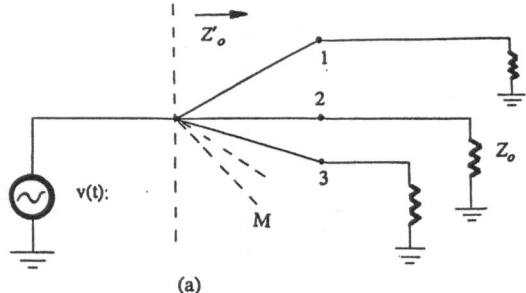

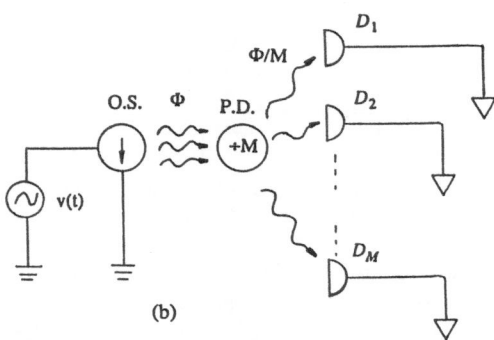

Fig. 3.6. Interconnection fan-out: (a) electrical, (b) optical. $v(t)$ is the driver for either system; Z_0': equivalent impedance of parallel lines; Z_0: impedance of individual lines; O.S.: optical source; P.D.: power divider; D_i: detectors; M: number of channels; Φ: flux generated by the optical source

A comparable ideal optical fan-out arrangement is illustrated in Fig. 3.6b. In this situation, the available optical flux Φ is split by an optical divider, directing Φ/M to each detector. Since each detector still requires the same charge to bring the gate to the threshold voltage, the optical power needed to drive M detectors also scales linearly with the number of detectors.

3.2.4 Ideal Optical Sources

The above results indicate that for both single and multiple fan-out conditions, the optical power required to charge a detector and an attached gate is comparable to the electrical power required for both gate-to-gate and chip-to-chip electrical connections. At the gate-to-gate level, the comparison was made with the power needed to drive the longest lines on the chip, and at this level of interconnection, the required power is a strong function of the length of the interconnection. The required power scales approximately linearly with the chip dimension. If smaller regions within the chip are considered, then the power can be considerably less. As an example, if a 1 mm cell within the chip is connected, the power needed is about 20 μW, or an order of magnitude less than the longest line requirement. These relationships point out that if the source were 100 % efficient, an optical interconnection would have a power advantage over the longest electrical connections on chips, which are 1 cm in length or larger.

At the chip-to-chip level, the power required for capacitive electrical connections is most strongly affected by the bonding pad capacitance. The capacitance of the wire length is small because of the large separation between the wire and the ground plane. In this case, the optical interconnect with an ideal source has a power advantage because the junction capacitance of the detector is significantly less than the bonding pad capacitance. The detector of the optical systems also requires significantly less power than an ideal terminated connection. For a given threshold voltage, the electrical power is fixed by the characteristic impedance of the line, and the line length does not affect the result.

These comparisons have considered only first-order electrical and optical effects. They have not taken line interference effects into account or the influence of improper line termination or losses in the transmission line. Although these comparisons seem to favor optical connections, they have not included the specific requirements of the optical source. To complete the analysis, the properties of semiconductor sources are considered in the next section.

3.2.5 Real Optical Sources

The previous discussion points out some requirements of the optical source for interconnect applications. In order to complete the comparison between optical and electrical interconnects, a more detailed investigation of these sources is presented in this section. Two sources which may potentially be integrated with the electronic logic are the semiconductor light emitting diode and laser diode. Although research into the development of each device is very active, the details of the semiconductor physics responsible for the operation of each are well documented [Refs. 3.10; 11, pp. 704–742; 12]. The purpose here is to establish the necessary source requirements and then to see how each device matches these requirements.

The parameters of interest for an optical interconnect source are the following:

1) Electrical-to-optical conversion efficiency η_T. This minimizes the power necessary for a transmission. For sources with an efficiency which differs from 100 %, the input–output power can be related by the simple expression

$$P_o = \eta_T V_F I_F , \tag{3.18}$$

where P_o is the optical output power, η_T the overall conversion efficiency, V_F the forward bias potential, and I_F the forward drive current.

2) Modulation properties. In order to be competitive with high transmission rate electrical systems, it should be possible to modulate the source at rates exceeding 1 GHz.

3) Required source area. The source must emit enough power to illuminate a detector with 200 μW of flux, and require less area than that needed by electrical transmission techniques.

4) Beam divergence angle. Ideally, this should be a controllable parameter

which will allow a full range of optical imaging elements to be considered. Emission angles, $\Delta\theta$, from 6° to 50° cover $f/10$ to $f/1$ imaging elements with a 1 cm separation between the imaging elements and the source.

5) Emission wavelength and spectral bandwidth. If wavelength-selective optical components are used, it is necessary to match the spectral output of the source with these elements. This is the case with optically recorded holographic materials. At present, the recording sensitivity of most of these materials is limited to a wavelength of about 780 nm or shorter. Therefore, ideally the source should have an emission wavelength below this value. It is also easier to design a high-resolution diffractive optical element to operate with a narrow band of reconstruction wavelengths. For this reason a narrow band source is desirable.

6) Sensitivity of the emission wavelength to changes in drive current, operating temperature, and fabrication tolerances. The imaging and efficiency properties of diffractive optical elements are strongly dependent on wavelength; therefore the source must provide a stable emission wavelength over its operating conditions.

The operating characteristics of LEDs and laser diodes are now evaluated with reference to these characteristics.

3.2.6 LED Emission Characteristics

A light emitting diode (LED) is a p-n junction capable of emitting light with the application of a forward biasing voltage. The forward bias injects minority carriers across the junction, producing a current. These injected carriers recombine randomly, producing photons. Light emerging from an LED is limited by: the conversion of electrons into photons and absorption of photons prior to leaving the diode; Fresnel reflection losses at the diode–air interface; and critical angle loss. Losses resulting from the internal conversion and absorption process are typically expressed in terms of an internal efficiency η_{in}, and those which occur when light leaves the diode, by an external efficiency η_{ext}. The optical power emitted by the diode can be expressed as

$$
\begin{aligned}
\Phi_{LED} &= \left[\frac{h\nu}{ec}\right] \eta_{ext}\,\eta_{in}\, I_F \\
&= \left[\frac{h\nu}{ec}\right] \eta_T\, I_F \\
&= \left[\frac{h\nu}{ec}\right] \eta_T\, J_F\, A_{LED} \;,
\end{aligned}
\tag{3.19}
$$

where $h\nu$ is the photon energy, e is the charge of an electron, c is the speed of light, I_F is the forward current, J_F is the forward current density, and A_{LED} is the junction area. This shows that the output power is linear with the current and the area of the emitter. The best efficiencies for direct gap LEDs are on the order of 10 %–15 %, and for indirect gap visible LEDs, a few tenths of a percent [Refs. 3.13; 14, p. 39]. A 10 % efficient source with a 1.4 V band gap can emit about 150 μW of optical power with 1 mW of applied electrical power.

The above value compares favorably with the power needed for a terminated electrical connection. However, the intensity emission pattern of a surface-emitting LED closely approximates a Lambertian profile with a $\cos \theta$ dependence, θ being the angle between the normal to the LED surface and the observation point. Therefore, the radiated power collected by either an imaging or a fiber interconnect system depends on the f/number of the collection element. For example, an $f/1$ collection lens receives about 20 %–25 % of the available power. Therefore, about 4 mW of electrical input power is needed to deliver 150 μW to a relatively low f/number system.

The area of the source needed to produce this flux is determined by its surface radiance L_e, and the current necessary to produce this radiance. The radiance of an emitting surface necessary to illuminate a lens with a given amount of flux is given by

$$L_e = \frac{\Phi_T}{A_{src} \, \Omega_L} \, , \tag{3.20}$$

where Φ_T is the flux which must be delivered to a receiver, A_{src} the emission area of the source, and Ω_L the solid angle of the collection lens. Therefore, a $(100 \, \mu m)^2$ area source illuminating an $f/1$ lens with 150 μW of flux must have a radiance of 2.5 W cm^{-2} sr^{-1}. An efficient Burrus-type double heterostructure LED [3.15] with 10 mA applied forward current is about 5 W cm^{-2} sr^{-1}. (The rated value of this LED is with 100 mA forward current, and it is assumed that the output characteristics can be scaled linearly to 10 mA.) If an $f/10$ collection lens is used however, the radiance must increase to 300 W cm^{-2} sr^{-1}. Therefore, LEDs are potentially useful only for low f/number systems.

The emission wavelength is fixed by the band gap of the materials forming the junction. Both direct-gap materials in the near-infrared and the infrared, and indirect-gap materials for emission in the visible have been developed [3.13]. In the direct-gap materials, a direct radiative recombination between the conduction and the valence band is allowed by momentum conservation. For indirect-gap materials, recombination takes place between energy levels of traps formed by impurity dopants. These dopants can be chosen so that the energy gap corresponds to a visible photon emission. The indirect process, however, is typically about two orders of magnitude less efficient than direct recombination.

The spectral bandwidth of the injection luminescence process is given by the approximate relation [Ref. 3.10, p. 236]

$$\frac{\Delta \lambda}{\lambda} = \frac{\Delta E_{ph}}{E_{ph}} \simeq \frac{2.5 kT}{E_{ph}} \, , \tag{3.21}$$

where k is the Boltzmann constant, T the temperature in Kelvin, and E_{ph} the photon energy. The bandwidth at $\lambda = 830$ nm is therefore about 36 nm.

The speed of an LED response at low drive currents is typically limited by the junction capacitance, and at high currents by the injected carrier lifetimes [Ref. 3.14, pp. 45–47], and the power radiated at modulation frequency f is given by

$$P(f) = \frac{P(0)}{[1 + (2\pi f \tau)^2]^{1/2}} \, , \tag{3.22}$$

where $P(0)$ is the radiated power at dc conditions, and τ the response time of the diode. Typical values for τ are from about 7 to 100 ns. Therefore, the maximum modulation bandwidth for an LED is a few hundred megahertz [Ref. 3.14, p. 41].

As discussed earlier, the optical source must deliver about 200 μW of optical power to a detector modulated at 1 GHz with input power comparable to that needed by an electrical system. In addition, the source size must be about the same as the electrical component it will replace (i.e. the 100 μm bonding pad), and the spectral properties of the emitted light compatible with spectral sensitivity of diffractive optical elements.

The properties of LEDs just outlined allow several conclusions to be drawn about their potential use as optical interconnects. First, if all of the light emitted from a 10 % efficient LED could be collected by an optical system, it would require a factor of 10 less power than a terminated chip-to-chip electrical connection. However, the large divergence angle reduces the available output power that can realistically be collected by even a low f/number lens. This makes the required LED power about equal to that of the terminated connection. Since the LED output power scales linearly with input power (3.19), and the terminated electrical fan-out power also scales linearly with the number of output connections, as seen in (3.17), both fan-out systems will require about the same amount of driving power. Second, the modulation capability of an LED is less than 1 GHz. A few hundred megahertz can be attained, but this requires a current of several hundred milliamps. In addition, the spectral bandwidth is broad. An image of this source formed by a diffractive imaging element will have a size that is larger than the source by a considerable factor [3.16]. It was also shown that when low f/number collection lenses are used, the radiance from an LED is adequate to allow 100 μm length emission surfaces to be used.

As indicated, the performance of an LED as an optical interconnect source is marginal. However, LEDs are easily fabricated and are readily made into surface emitters. These attributes might make them attractive for lower speed interconnect systems.

3.2.7 Laser Diode Characteristics

A laser diode is formed by combining the effects of stimulated emission and optical feedback to produce a high efficiency, directional optical source [3.12]. A threshold current must be applied to create a population inversion and to overcome the losses of the system. The threshold current density is given by

$$J_{th} = \frac{J_0 d}{\eta} + \frac{J_0 d}{g_0 \eta \Gamma} \left[\alpha + \frac{1}{L_t} \ln \left(\frac{1}{R} \right) \right] , \tag{3.23}$$

where Γ is the confinement factor of the light in the active layer relative to light both within and outside the active layer, α is the loss per unit length from free carrier absorption and defect-center scattering, L_t is the length of the cavity, R the reflectance of the cavity ends, η the quantum efficiency, and d the active layer thickness. J_0 and g_0 are empirical terms derived from fitting a linear curve to the gain versus current relation for a particular material.

Recently it has been demonstrated that in optimized stripe and double heterostructure lasers, the threshold level can be reduced to below 1 mA [3.17]. For these structures, it may be possible to reduce this by another factor of two and still produce the needed population inversion [3.18]. External differential quantum efficiencies on the order of 50 %–60 % have been achieved for these devices [3.18]. The differential quantum efficiency η_D is defined as the conversion of electrical to optical power after achieving threshold

$$\eta_D = \frac{e}{\epsilon_{ph}} \frac{\Delta \Phi}{\Delta I} , \tag{3.24}$$

or,

$$\Delta \Phi = \left(\frac{\eta_D \epsilon_{ph}}{e} \right) \Delta I$$

where ϵ_{ph} is the energy of an emitted photon.

This relation, the threshold current, and the forward biasing potential for the junction allow the optical output power to be calculated as a function of the electrical input power. For a GaAlAs laser diode, the forward potential drop is 1.83 V. A set of curves is shown in Fig. 3.7 for: a) a threshold current $i_{th} = 1$ mA, $\eta_D = 0.5$, and an emission wavelength $\lambda_e = 830$ nm; b) $i_{th} = 1$ mA, $\lambda_e = 700$ nm, $\eta_D = 0.5$; and c) $i_{th} = 10$ mA, $\lambda_e = 830$ nm, $\eta_D = 0.5$. These calculations show that the threshold current is the largest factor in determining the required operation power. Reducing the emission wavelength also reduces the necessary power. For case (a), 3.5 mW input power produces 800 μW optical power. This is less than the ideal terminated line requirement by almost a factor of three. This available optical flux can power four detectors, which means that, for these conditions, an optical system can provide a fan-out of four, with one-third the power of a single terminated connection. This is a definite power advantage over electrical connections.

The modulation properties are another consideration for the laser diode. The most serious limitation of direct current modulation is resonance effects occurring between the energy stored in the injected minority carriers and the optical field in the cavity [3.19]. These difficulties can be controlled by designing the cavity to enhance carrier confinement, allowing modulation rates of several gigahertz [3.20].

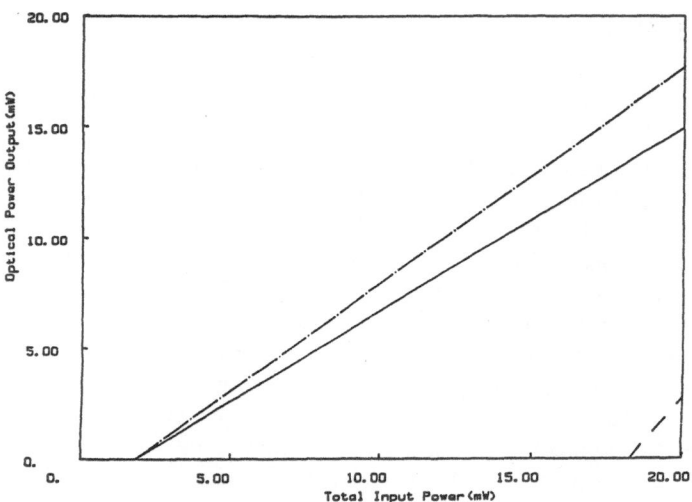

Fig. 3.7. Laser diode output power [mW] vs input power [mW]; (—) i_{th} = 1.0 mA, η_D = 0.5, λ_e = 830 nm; (-----) i_{th} = 1.0 mA, η_D = 0.5, λ_e = 700 nm; (- - - -)i_{th} = 10 mA, η_D = 0.5, λ_e = 830 nm

As mentioned earlier, other appropriate features which must be considered are the emission wavelength and spectral bandwidth, the size, and the beam divergence angle. The emission wavelength is fixed by the spectral bandwidth of the gain medium and the cavity mode spacing. GaAlAs double heterostructure and multiple quantum well lasers have been fabricated with emission wavelengths from 750 nm to 850 nm and linewidths of a few tenths of a nanometer. Typical dimensions for room temperature lasers are 200–250 μm × 1 − 10 μm × 0.1 − 0.4 μm.

The geometry of the cavity also determines the form of the intensity pattern emitted from the laser. The intensity distribution as a function of angle normalized to the intensity at $\theta = 0°$ is [3.21]

$$\frac{I(\theta)}{I(0)} = \frac{\cos^2(\theta)|\int_{-\infty}^{\infty} E_y(x,0)\exp[i\sin(\theta)k_0 x]dx|^2}{|\int_{-\infty}^{\infty} E_y(x,0)dx|^2} , \qquad (3.25)$$

where $E_y(x,0)$ is the electric field amplitude at the exiting surface of the junction and $\cos \theta$ is the diffraction obliquity factor. Usually, the cross section of the junction is not symmetrical and makes the divergence angle different along each axis of the emitting face. The divergence angles for a typical double heterostructure laser with a junction 0.2 μm × 12 μm are 45° by 10° [Ref. 3.11, p. 761]. The asymmetric emission characteristics are difficult design constraints for imaging systems, and it would be better if the beam were collimated or had a symmetric divergence angle of a few degrees.

Laser diodes are typically edge or horizontal emitters; however, many interconnect applications considered require surface or off-plane directed emission.

One technique for circumventing this difficulty employs an etched reflector at 45° which deflects the output from an edge emitting laser diode vertically. A demonstration of this technique has recently been reported in which the individual outputs from 16 lasers in a two-dimensional array were made to emit vertically by using 45° reflectors [3.22].

This comparison indicates that the power requirements, modulation characteristics, and spectral emission properties of laser diodes satisfy the needs for an optical interconnect system. In addition, it has been shown that there is a significant power advantage over both single and multiple fan-out terminated electrical connections. The size and divergence angle of the diodes, however, are not optimum. The length is about a factor of two greater than the design goal of 100 μm; however, the width is about 10 μm. This might allow two lasers to be mounted within a 200 μm × 200 μm area. The divergence angle is a more serious problem. If existing layer cavity designs are used, additional beam shaping will be necessary.

3.3 Optical Distribution Techniques

An important design consideration of an optical interconnect system is the method of distributing the optical signal. This can be accomplished by using an index guiding medium or an imaging element to divide the signal into a desired interconnect pattern. The advantages and limitations of these approaches are the subject of this section.

3.3.1 Guided Optical Distribution Techniques

Two candidate guided wave materials are optical fibers and integrated optical waveguides. Both of these components have been extensively studied for application to optical communication systems.

Fiber optic waveguides are low loss transmission media with essentially unlimited bandwidth over lengths of several centimeters. Typical losses for plastic fibers are 0.01 dB/cm, and for glass fibers, 10^{-5} dB/cm. With a typical optical carrier frequency of 2×10^{14} Hz, many 1–10 Gb/s transmission bands are accommodated. Another useful aspect of fiber waveguides for interconnect applications is that they can be used to connect different geometrical planes. This feature has potentially useful applications for replacing back-plane connections between boards of a computer system.

One disadvantage of fiber systems for interconnect applications has been their limited fan-out capability. This restricted potential applications to external-to-internal point-to-point connections, which did not provide a significant advantage over electrical connection techniques. However, a recently developed fiber coupling device has been demonstrated that has relatively high fan-out with very low coupling loss [3.23]. This technique has given a fan-out of 40 with a coupling loss of 0.5 dB.

A major difficulty of fiber interconnect systems is the need to make individual fiber connections to detectors. Several techniques for attaching fibers to integrated circuits use V-groove alignment fixtures [3.24], a tapered fiber coupler [3.25], and a lateral fiber coupling technique [3.26]. Alignment of these devices is critical, and interconnect spacing is limited by the fiber and cladding diameters.

Integrated optics is a second approach to guided optical interconnects. Specific advantages of integrated optics over fibers and free space imaging elements are: the interconnect patterns can be formed by lithographic techniques; optical waveguides and active switching elements can be formed directly in GaAs substrates along with sources and detectors; and, the interconnect system is relatively rugged. Although such devices are restricted to a plane, they may have several useful applications. For example, 1 : N star couplers have been fabricated with N on the order of 100. These devices, along with tapered waveguide couplers, can be used to distribute optical signals at the chip level. Polymer waveguides have also been fabricated for distributing signals at the board and back-plane interconnect levels [3.27].

3.3.2 Free Space Distribution Methods

Optical information can also be distributed to processing locations by free space propagation techniques. These include the use of imaging and nonimaging lenses and mirrors, and special purpose holographic optical elements. One of the primary

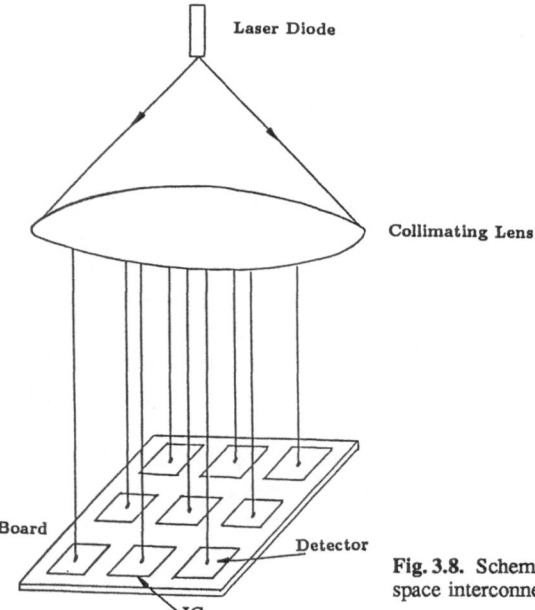

Fig. 3.8. Schematic representation of an unfocused free space interconnect system

advantages of this approach is elimination of planar constraints of the interconnects. The principles of physical optics govern the propagation, and since photons can pass freely through each other, optical channels are allowed to cross.

Free space interconnects can be categorized into focused and unfocused systems. A simple example of an unfocused system is illustrated in Fig. 3.8. This arrangement represents an external-to-internal global interconnect scheme. Light from a source is collimated by a lens and then illuminates the receiving area. The optical path lengths from the source to each receiver position are equal, therefore signals are received without relative delays or skew. This is an important consideration for clock signal distribution. However, if the receiving area is not completely occupied with detectors, the method is not efficient. A possible solution to this problem is to construct a multi-level circuit with a matrix of integrated detectors on the top layer, a layer of opto-receivers, and finally the logic electronics, connected by vertical vias.

An alternative free space propagation method uses focusing or imaging elements to concentrate light in a number of receiver positions. This method is more efficient at distributing optical power and reduces the requirements on the opto-receiver.

3.4 First-Order Analysis of Holographic Interconnects

In this section, the geometrical design parameters of a free space holographic optical element are evaluated. The intended function of the hologram is to connect two or more VLSI integrated circuits at the chip-to-chip connection level.

3.4.1 Interconnect Parameters

In order to connect several VLSI integrated circuits, the hologram must provide several imaging functions. First, several hundred independent transmission paths must be formed. This is required in order to take advantage of the computational power of an integrated circuit with a high density of logic gates. The interconnects must also have the potential of connecting one transmission point to more than one receiver. This fan-out feature determines the parallel processing capability of the system of integrated circuits. Greater fan-out implies that a signal can be transferred to many locations simultaneously or with a high degree of parallelism.

Electrical connections at the chip-to-chip level are formed by thermally or ultrasonically connecting wires to bonding pads. The minimum lengths of these pads are on the order of 100 μm, which is limited by the wire bonding technique. These pads are separated by distances that are also on the order of 100 μm. For design purposes, the bonding pad dimension can be used as the minimum image resolution requirement.

Adjacent integrated circuits are separated by distances ranging from a few millimeters to several centimeters, depending on the width of the integrated cir-

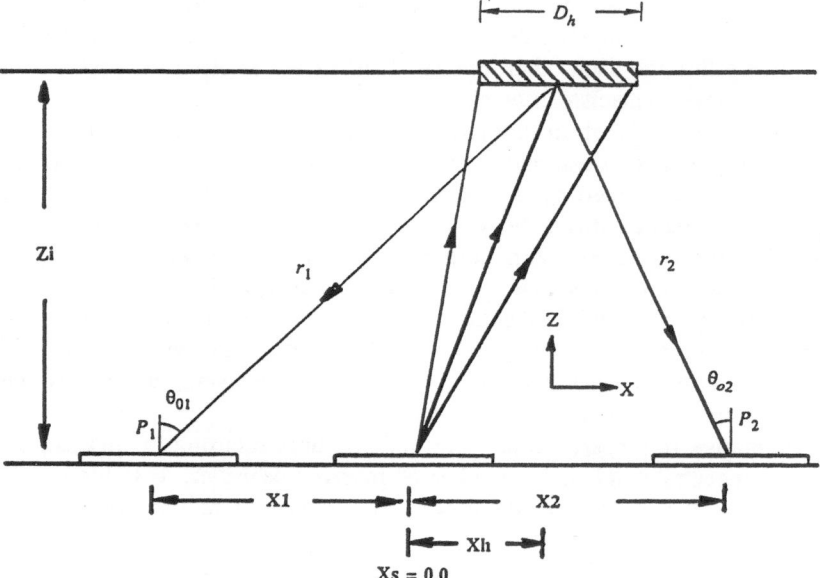

Fig. 3.9. Schematic representation of a focused free space interconnect with a fan-out of two. X_s: source position; X_1, X_2: detector positions; D_h: the hologram position; θ_{o1}, θ_{o2}: angles between the receiving ICs and the hologram

cuit and the size of the board on which the circuits are mounted. These distances, and the height of the hologram above this substrate, determine the field-of-view of the hologram.

The case of imaging to two detectors is illustrated in Fig. 3.9. This represents a single optical channel with a fan-out of $M = 2$. The hologram is located at a distance Z_i from the plane on which the ICs are mounted and has an aperture with diameter D_h. The two receiving ICs are located at distances X_1 and X_2 from the optical source. For convenience, the source is assumed to be located at a position $X_s = 0.0$.

An optimum hologram design can be obtained by maximizing the number of channels M_c and the difference in detector positions $(X_1 - X_2)$, minimizing the hologram–substrate separation Z_i and the hologram diameter D_h, and maintaining an acceptable image resolution at each detector $(\Delta d_{1,2})$. The maximum detector separation (ΔX) at which acceptable image resolution is maintained can be used as a figure of merit for comparing different hologram designs. Using the bonding pad length ($100\,\mu$m) as an acceptable resolution limit, this merit function has the form

$$\Delta X = (X_1 - X_2) = f(D_h, Z_i, M, \Delta d_{1,2} = 100\,\mu\text{m}) . \tag{3.26}$$

The specific functional relationship is determined by the way in which the image is reconstructed.

Two methods of illuminating the volume hologram are considered. In the first configuration, the hologram is illuminated at the Bragg angle, while in the second it is illuminated at angles that are displaced from this condition. When a hologram is reconstructed at angles which differ from the formation condition, aberrations result. An exact description of an aberrated image determined by diffraction analysis can be complicated [3.28]. However, a reasonable approximation for the image resolution can be obtained by assuming that the spot size is equal to the sum of the width Δd_{dj} of the diffraction-limited image plus the width Δd_{aj} of the aberrated image predicted by ray tracing. The general form of the spot size can then be expressed as

$$\Delta d_j \simeq \Delta d_{dj} + \Delta d_{aj} \ . \tag{3.27}$$

This approximation places a constraint on the minimum size of the hologram aperture in a known reconstruction configuration, and will be used for calculating the image resolution for the merit function.

When the hologram is illuminated at the Bragg angle [3.29], $\Delta d_{aj} = 0$. For the case of a circular aperture and a converging spherical beam

$$\Delta d_j = \frac{2.44\lambda r_j}{D_h \cos^2 \theta_{oj}} \ , \quad j = 1,2 \ . \tag{3.28}$$

X_1 and X_2 are related to the angles θ_{o1} and θ_{o2} by

$$X_1 - X_h = Z_i \tan \theta_{o1} \ ,$$
$$X_2 - X_h = Z_i \tan \theta_{o2} \ ,$$

where θ_{oj} is the angle between the normal and the hologram aperture and a ray to the point of focus for one of the images. One $\cos \theta_{oj}$ factor is due to the reduced effective aperture and the second to the projection of the diffraction pattern onto the substrate. The radial distance r_j is equal to

$$r_j = \frac{Z_i}{\cos \theta_{oj}} \ .$$

If the holographic element is illuminated at an angle that is different from the construction angle, the reconstruction wavefront will be distorted and aberrations result. For this situation both Δd_{dj} and Δd_{aj} will generally have nonzero values. The approximate size of the aberration term Δd_{aj} can be estimated by tracing several rays to the substrate plane and determining the size of the spot on this plane. Rays are traced by using the grating vector equation

$$l_{1q} = \frac{K_{xq}}{2\pi/\lambda} + l_{rq} \ , \tag{3.29}$$

Where l_{1q} is the direction cosine to the image at P_1, l_{rq} is the direction cosine of the reconstruction ray from a point on the transmitting IC at the position Y_r to the point q on the hologram, and K_{xq} is the x component of the volume

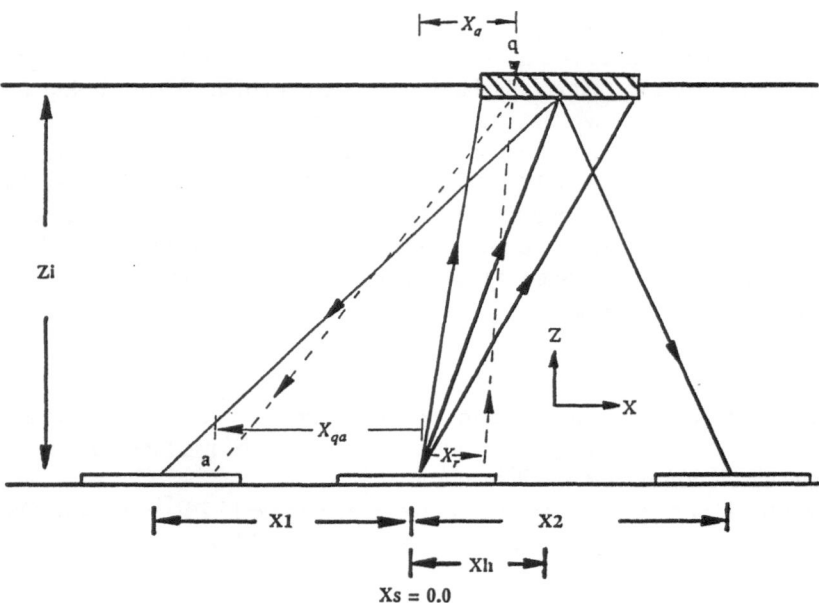

Fig. 3.10. Hologram and image coordinates used in calculating the image aberrations

grating vector along the surface of the hologram at the point q on the hologram aperture (Fig. 3.10). The spot size of the aberrated image is determined from the difference in distance between the maximum (point a) and minimum (point b) intercepts on the substrate plane:

$$x_{qa} = x_q + r_{qa}l_{qa} , \tag{3.30}$$

with

$$r_{qa} = \frac{Z_i}{(1 - l_{qa}^2)^{1/2}} . \tag{3.31}$$

A similar expression exists for the minimum intercept x_{pb}. The difference between these positions gives the size of the image along the x axis, and can be expressed as

$$\Delta d_{aj} = x_{qa} - x_{pb} = (x_q - x_p) + (l_{jqa}r_{jqa} - l_{jpb}r_{jpb}) , \tag{3.32}$$

where l_{jqa} refers to the direction cosine from point q on the hologram to point a on the detector j. In this case, the possible detector positions X_1 and X_2 are also functions of the direction cosines, K_{xq}, and the distances r_{jqa}.

The complete expression for the spot size becomes

$$\Delta d_j = \Delta d_{dj} + \Delta \dot{d}_{aj}$$
$$= (x_q - x_p) + (l_{jqa}r_{jqa} - l_{jpb}r_{jpb}) + \frac{2.44\lambda r_j}{D_h \cos^2 \theta_{oj}} , \quad j = 1,2 . \tag{3.33}$$

3.4.2 Interconnect Designs

The above resolution and spatial relationships are now used to evaluate two different hologram designs. The first is a multifaceted hologram recorded with a space-variant set of object beams for each facet. The second hologram is a single aperture multiple-focusing element which is easier to fabricate, but has a spatially-invariant set of object points.

(a) Multifacet Hologram. For this arrangement, the hologram aperture is divided into a number of subholograms or facets [3.30]. Each facet is a multiple-image hologram, which is reconstructed with the conjugate of the reference wave. When the reference beam is a converging spherical field and the object waves diverging spherical fields, a diverging reconstruction beam brings the various object beams to focus as shown in Fig. 3.11a. For this hologram, a separate source illuminates each facet, and it is assumed that the spatial output pattern of each laser diode used in the reconstruction illuminates a single facet. Each subhologram can then be encoded with a unique combination of object beams, which allow a distinct set of locations on an integrated circuit to be addressed. The number of interconnect channels is determined by the number of facets that can be formed on the available hologram aperture, while the minimum facet diameter is constrained by the requirement to maintain a desired image resolution spot size at each detector position. The resolution in this case is determined from (3.28).

(b) Single-Aperture Design. The second design is a single multiple grating hologram. This element is formed with the reference beam propagating from

A) Faceted Hologram Design:

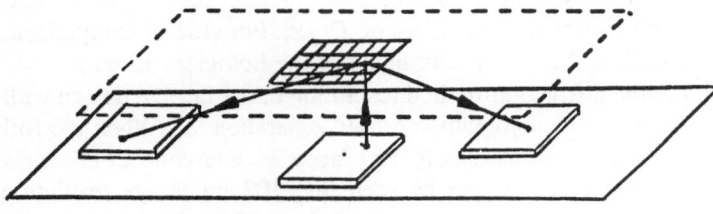

B) Full Aperture Multiple Grating Element:

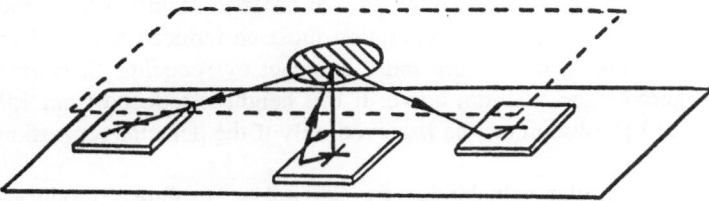

Fig. 3.11. Multiple image hologram designs: (a) faceted hologram design; (b) single-aperture design

83

the center of the transmitting IC and the object beams from the centers of the receiving integrated circuits as shown in Fig. 3.11b. During reconstruction, a source illuminates the hologram from a position displaced from the center of the IC. The image resolution in this case is given by (3.33).

This element is simpler to fabricate than the multifacet design; however, the flexibility of arbitrarily positioning the receivers is lost. In addition, since this hologram is not reconstructed with the conjugate of the reference beam, the images are aberrated. The maximum distance that a source can be displaced from the center of the transmitting IC is determined by the separation at which the aberrated image equals the detector diameter. The number of interconnect channels equals the number of sources that can be mounted within the area defined by this radius. Since this comparison is primarily concerned with the performance of the imaging element, the difficulties associated with fabricating a large number of sources in a small region of an integrated circuit is not addressed. It is assumed that the number of sources is limited by the total available substrate area divided by the area required by a source. For this comparison, the area required by a source is assumed to be the same as for a bonding pad.

3.4.3 Comparison of Designs

These designs provide trade-offs between fabrication complexity, and image resolution and flexibility. These parameters can be quantified by using (3.28) and (3.33) to compute the range of detector separations (X_1 and X_2) that can be addressed with the required image resolution. The results can be expressed in terms of the hologram aperture D_h, and the hologram height above the substrate Z_i.

To illustrate the dependence of the image resolution on as many of these parameters as possible, the logarithm of the focused spot size, $\log(\Delta d_j)$ is plotted as a function of X_1 and X_2 for several values of D_h/Z_i. For ease of comparison, the value for X_s as well as the center coordinate of the hologram is zero.

Figure 3.12 shows the diffraction-limited resolution of the faceted design with three ratios of diameter D_h to hologram-substrate separation Z_i. When the full aperture diameter is 1.0 cm, approximately 100 facets or interconnect channels can be formed. From the curve, it can be seen that 100 μm image resolution can be achieved with a detector separation up to 3.2 cm. If the facet diameter is reduced by a factor of 3, the number of interconnect channels can be increased to 400. The dashed curve in Fig. 3.12 shows that in order to maintain the same resolution at both images, the detector separation must be reduced from 3.2 to 2.2 cm. The result of making the system more compact by reducing Z_i is also indicated in this figure by the dash-dot curve. If this height is reduced from 1.0 to 0.5 cm, the required resolution can be achieved only if the detector separation is reduced to 2.2 cm.

Figure 3.13 shows similar calculations for the image resolution produced with the single-aperture element. In this case, the ratio D_h/Z_i was optimized to

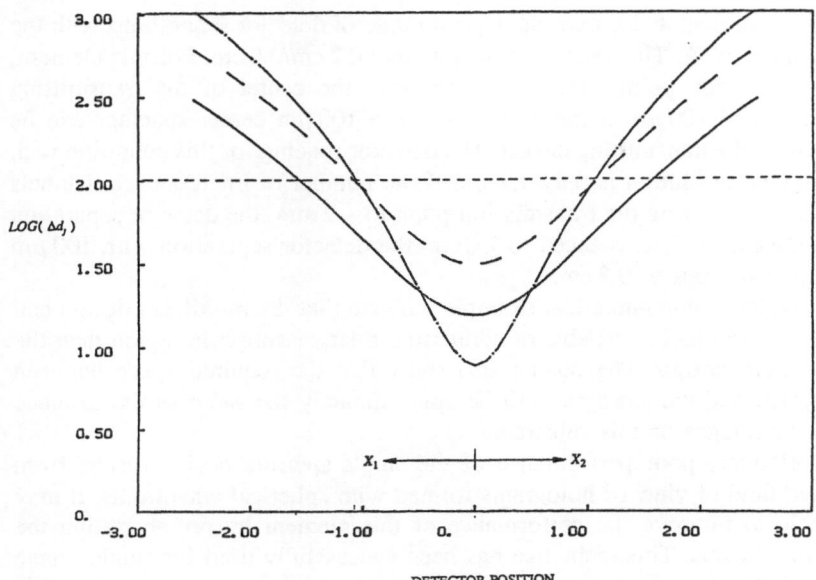

Fig. 3.12. Faceted hologram image resolution as a function of the detector separation from the source. X_1 and X_2 are assumed to be on opposite sides of the source at $X = 0.0$: (——) $D_h Z_i = 0.1 \, \text{cm}/1.0 \, \text{cm}$; (- - - -) $D_h/Z_i = 0.05 \, \text{cm}/1.0 \, \text{cm}$; (-·-·-) $D_h/Z_i = 0.1 \, \text{cm}/0.5 \, \text{cm}$; the horizontal dashed curve shows $100 \, \mu\text{m}$ resolution

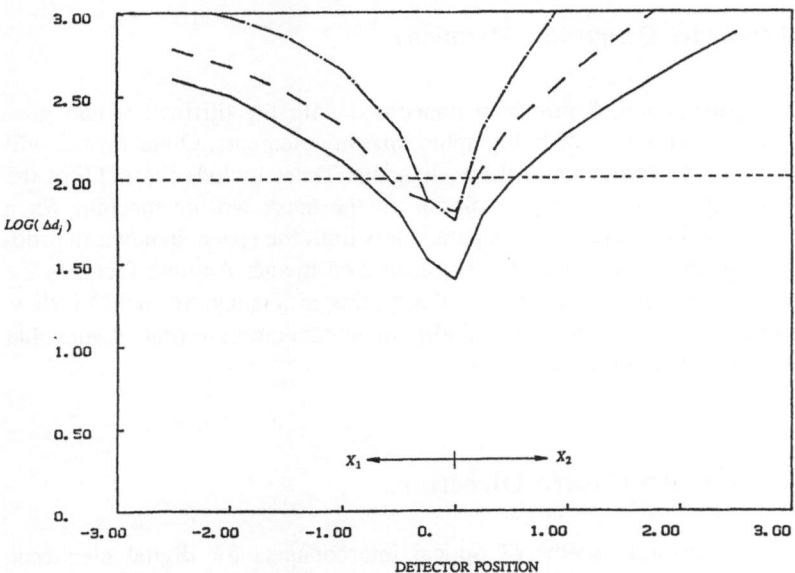

Fig. 3.13. Single-aperture hologram image resolution as a function of detector separation from the source. X_1 and X_2 are assumed to be on opposite sides of the source at $X = 0.0$: $D_h/Z_i = (0.2 \, \text{cm})/(2.0 \, \text{cm})$; (——) $X_r = -1 \, \text{mm}$; (- - - -) $X_r = -2 \, \text{mm}$ (-·-·-) $D_h/Z_i = (0.2 \, \text{cm})/(1.0 \, \text{cm})$, $X_r = -1 \, \text{mm}$; the horizontal dashed curve shows $100 \, \mu\text{m}$ resolution

form the smallest spot size over the largest range of detector separations with the smallest value of Z_i. This ratio is approximately 0.2 cm/2.0 cm. For this element, at a reconstruction point located -1 mm from the center of the transmitting IC, one hundred 100 μm diameter sources with 100 μm center spacings can be positioned on the transmitting circuit. The detector spacing for this condition with 100 μm image resolution is only 1.2 cm. If the number of interconnect channels is increased by moving the transmission point to -2 mm, the detector separation drops to 0.8 cm. If Z_i is reduced to 1.0 cm, the detector separation with 100 μm resolution also drops to 0.8 cm.

The results of this numerical example indicate that the multifacet design can be more compact and is capable of addressing a larger substrate region than the single-aperture design. The results also show that the required space between the hologram and the substrate will be approximately the same as the distance between the images on this substrate.

The relatively poor performance of the single-aperture design results from the limited field of view of holograms formed with spherical wavefronts. It may be possible to improve the performance of this element by pre-aberrating the construction beams. This technique has been successfully used for single image elements [3.31], but has not yet been applied to multiple image holograms. The drawback of using the multifacet design is the requirement to encode a distinct pattern of object beams in each subhologram as well as the need to have highly directive sources.

3.4.4 Higher-Order Design Considerations

The previous discussion is primarily concerned with the diffraction and geometrical imaging properties of holographic imaging elements. Other factors will also influence the performance of these elements. These include the MTF of the optical recording material or the resolution of the mask writing machine for a computer generated hologram. These parameters limit the space–bandwidth product and the angular separation of the reconstructed image. Another factor is the hologram thickness, which affects both the grating efficiency and field-of-view. The hologram mounting technique and alignment accuracy are other issues that must be included in a complete design.

3.5 Summary and Future Directions

In this chapter, several aspects of optical interconnects for digital electronic computer systems were examined. A comparison of electrical and optical systems showed that an optical transmission system at the chip-to-chip connection level with even low levels of fan-out requires less power than an electrical system. An analysis of free space holographic imaging elements then revealed that there is a natural restriction on the minimum size and position of this element due to its diffraction and aberration characteristics.

Several important issues must still be resolved, however, before practical systems can be realized. If each communication point requires a receiver and transmitter, then several hundred opto-receivers and laser diodes will have to be fabricated on an integrated circuit. This is a difficult problem for several reasons. First, if silicon logic elements are used, then GaAs or some other direct band gap material will have to be grown on silicon, or packaged in a hybrid form to make light emitting devices. However, significant advances must be made before a large number of lasers, detectors, and logic elements can be integrated. This integration process must also isolate electrical and optical interference effects which result from driving laser diodes at high modulation frequencies. Another requirement for the previous design is a vertical emitting laser diode. As mentioned in Sect. 3.2.7, work is progressing in this area, however, it is still in the preliminary stages of development.

These difficulties imply that, at the present time, interconnect designs which require a large number of source-detector pairs are not practical. One important application which does not have such a requirement, however, is the distribution of a synchronous optical clock signal. Implementing this system requires a single source which does not have to be integrated with the logic elements, and possibly only a single opto-receiver, integrated on a chip. These components can be fabricated with available device technologies.

Another approach to realizing an interconnect system with free space elements is to use a single source in conjunction with a light modulator. Several designs have been presented using this concept [3.37, 38], and considerable research effort is going on toward developing large arrays of high speed spatial light modulators [3.34, 35].

Recently, a new concept for free-space imaging interconnects has appeared which offers a number of advantages over conventional holographic interconnects [3.36–40]. These elements combine a totally internally reflected (TIR) beam with one or more holographic optical elements. An illustration of one form of this device is shown in Fig. 3.14. Light enters the hologram normal to the surface,

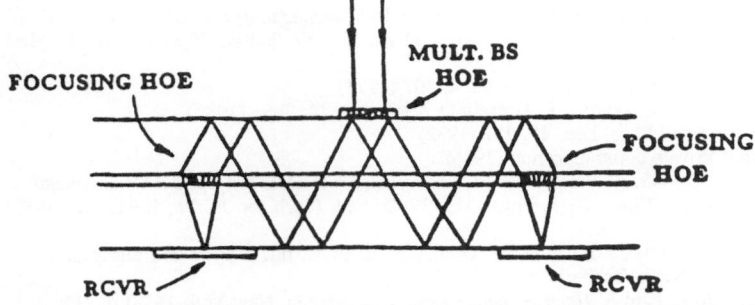

Fig. 3.14. Holographic interconnect which combines one or more holographic input and output couplers with a totally internally reflected beam. In this case, the input coupler is a multiple grating, and the output coupler is embedded within the guiding substrate. HOE: holographic optical element; RCVR: receiver

diffracts from a transmission hologram, and propagates through the substrate as a TIR beam. When at the location of a receiver, an output coupling hologram focuses light to a detector. The first benefit of this arrangement is that the beam can be translated over relatively large distances without the need to increase the height of the hologram above the detector plane. Next, the set of holograms can be packaged in a monolithic block, which helps to reduce alignment and mounting problems. Also, since pairs of holograms are used, some degree of chromatic compensation for laser diode wavelength variation can be realized [3.50]. Finally, the arrangement allows entire images to be transferred through the system, which maintains the parallel information character of the optical system. The holographic optical elements in these devices can be formed by either optical or computer generated methods, providing a large degree of flexibility in the fabrication process.

In conclusion, it appears that several important interconnect problems are amenable to a solution with available optical interconnect technologies. Solution of the engineering problems associated with these systems should add a new dimension to the performance of digital electronic computer systems.

References

3.1 H. Tajima, Y. Okada, K. Tamura: Trans. Inst. Electron. Commun. Eng. Jpn. E **66**, 47 (1983)
3.2 M. Hatamian, L.A. Hornak, T.E. Little, S.K. Tewksbury, P. Franzon: AT&T Tech. J. **66**, 13 (1987)
3.3 R. McRuer, J.P. Wilde, L. Hesselink, J.W. Goodman: SPIE O-E LAE **881-31**, 89 (1988)
3.4 C. Mead, L. Conway: *Introduction to VLSI Systems* (Addison-Wesley, Reading, MA 1980)
3.5 K.C. Saraswat, F. Mohammadi: IEEE Trans. ED-**29**, 645 (1982)
3.6 P.M. Solomon: Proc. IEEE **70**, 489 (1982)
3.7 A. Barna: *High Speed Pulse and Digital Techniques* (Wiley, New York 1980) p. 143
3.8 S.E. Miller, A.G. Chynoweth: *Optical Fiber Telecommunications* (Academic, New York 1979)
3.9 G. Luckovsky, R.B. Emmons: Appl. Opt. **4**, 697 (1965)
3.10 J. Gowar: *Optical Communication Systems* (Prentice Hall, Englewood Cliffs, NJ 1984)
3.11 S.M. Sze: *Physics of Semiconductor Devices*, 2nd ed. (Wiley, New York 1981)
3.12 A. Yariv: *Introduction to Optical Electronics* (Holt, Rhinehart, and Winston, New York 1971) pp. 160-166
3.13 A.A. Bergh, P.J. Dean: Proc. IEEEE **60**, 156 (1972)
3.14 H. Kressel, M. Ettenberg, J.P. Wittke, I. Ladany: In *Semiconductor Devices for Optical Communication*, ed. by H. Kressel, Topics Appl. Phys., Vol. 39, 2nd ed. (Springer, Berlin, Heidelberg 1982)
3.15 C.A. Burrus, B.I. Miller: Opt. Commun. **4**, 307 (1971)
3.16 R.K. Kostuk, J.W. Goodman, L. Hesselink: Appl. Opt. **24**, 2851 (1985)
3.17 W.T. Tsang: Appl. Phys. Lett. **40**, 217 (1982)
3.18 W. Streifer: Private communication (1986)
3.19 G. Arnold, P. Russer, K. Petermann: In *Semiconductor Devices for Optical Communication*, ed. by H. Kressel, Topics Appl. Phys., Vol. 39, 2nd ed. (Springer, Berlin, Heidelberg 1982) pp. 213-242
3.20 M. Maeda, K. Nagano, I. Ikuhima, M. Tanaka, K. Saito, R. Ito: In Proc. 3rd European Conf. Opt. Commun. NTG-Fachberichte **59**, 120 (1977)
3.21 H.C. Casey, M.P. Panish: *Heterostructure Lasers* (Academic, New York 1978) p. 71
3.22 J.N. Walpole, Z.L. Liau: Appl. Phys. Lett. **48**, 1633 (1986)
3.23 R. Arrathoon: SPIE O-E LASE **898-06**, 121 (1988)
3.24 P.R. Haugen, S. Rychnovsky, A. Husain, L.D. Hutcheson: Opt Eng. **25**, 1076 (1986)

3.25   P.R. Prucnal, E.R. Fossum, R.M. Osgood: Proc. SPIE **625**, 154 (1986)

3.26   D.H. Hartman: Opt. Eng. **25**, 1086 (1986)

3.27   C.T. Sullivan, A. Husain: SPIE O-E LASE **881-27**, 88 (1988)

3.28   M. Born, E. Wolf: *Principles of Optics* (Pergamon, Oxford 1975) pp. 459–490

3.29   H. Kogelnik: Bell Syst. Tech. J. **48**, 2909 (1969)

3.30   S.K. Case, P.R. Haugen, O.J. Lokberg: Appl. Opt. **20**, 2670 (1981)

3.31   J.N. Cederquist, J.R. Fienup: J. Opt. Soc. Am. **77**, 699 (1987)

3.32   J. Taboury, J.M. Wang, P. Chavel, F. Devos: Topical Meet. Opt. Comput. Tech. Digest **11**, 31 (1987)

3.33   A.A. Sawchuk, B.K. Jenkins: Proc. SPIE **625**, 143 (1986)

3.34   A.D. Fisher: Topical Meet. Opt. Comput. Tech. Digest TuC1-1 (1985)

3.35   S.H. Lee, S.C. Esner, M.A. Title, T.J. Drabik: Opt. Eng. **25**, 250 (1986)

3.36   T. Jannson, S.-H. Lin: OSA Topical Meet. Spatial Light Modulators and Appl. **8**, 56 (1988)

3.37   K.-H. Brenner, F. Sauer: Appl. Opt. **27**, 4251 (1988)

3.38   F. Sauer: Appl. Opt. **28**, 386 (1989)

3.39   J. Jahns, A. Huang: Appl. Opt. **28**, 1602 (1989)

3.40   R.K. Kostuk, Y.-T. Huang, D. Hetherington, M. Kato: Appl. Opt., Vol. 28, 4939 (1989)

4. First Implementations of Optical Digital Computing Circuits Using Nonlinear Devices

A.C. Walker, B.S. Wherrett, and S.D. Smith

With 43 Figures

The demonstration, during the late 1970s, of optical bistability in micrometer-dimension semiconductor nonlinear etalons stimulated great interest in the possibility of using devices of this type as digital optical processing elements. Particularly exciting was the prospect of massively parallel optical computers based around two-dimensional arrays of such optically nonlinear devices. This chapter describes some first, very basic, steps towards this ultimate goal. It covers the operational characteristics of bistable optical cavities, the techniques by which these are best exploited as binary logic elements, and the various digital optical circuits and prototype processing architectures that have been operated experimentally.

4.1 Introduction

4.1.1 Optical Computing Architecture Considerations

Over the past 40 years the computing power of electronic machines has risen from a few hundred operations per second to above 10^9. While this seven or eight orders-of-magnitude improvement has, to some extent, been achieved by reducing the gate delay times of individual switches (three orders of magnitude), it has mainly been the result of an increase in processing *parallelism*. Such parallelism has taken many forms. The first step was away from the bit-serial computation to bit parallelism. This was followed by linear pipelining, in which a sequence of processing units pass on information in a manner analogous to a factory production line; then functional parallelism, in which different units simultaneously address complementary parts of a problem; and finally, linked microprocessors or computers. The greatest parallelism, and for the ideally matched task the greatest computing power, is achieved by large arrays of relatively simple processing cells working in lockstep. The first commercial machine built on these lines was the ICL DAP (Distributed Array Processor) with 64×64 processors each linked to its nearest neighbors [4.1]. Smaller, more compact versions are now made by AMT (Active Memory Technology, Reading, UK, and Irvine, Calif.). The larger and very powerful CM-2 (Connection Machine) has 65 536 processors, connected in a form of hypercube [4.2]; this machine is made by TM (Thinking Machines, Cambridge, Mass.). Recent similar machines, not commercially available, are the Goodyear MPP (Massively Parallel Processor) [4.3], and the

Springer Series in Electronics and Photonics, Vol. 30
Nonlinear Photonics Editors: H.M. Gibbs · G. Khitrova · N. Peyghambarian
© Springer-Verlag Berlin Heidelberg 1990

CLIP (Cellular Logic Image Processor) from University College, London [4.4]. The MPP has an array of 128×128 bit-processing elements, each linked to its nearest neighbors, while the CLIP has a similar number of processors but with much less memory than the other machines. Parallelism in these machines is restricted to on-chip architecture. The number of pins available for chip-to-chip interconnection remains on the order of 100, greatly limiting data throughput rates. The MPP load and down-load times, for example, exceed milliseconds.

It is in the context of highly parallel spatially invariant interconnections that optics has much to offer. The fundamental factor behind much of the interest in optical computing is the enormous information handling capacity of simple optical imaging systems. For example, a well-corrected lens is easily capable of resolving 10^6 points in an input plane and connecting them in parallel to a corresponding set of points in the image plane. To determine the overall data throughput rate, we can assume, conservatively, that each channel could be modulated on a picosecond time scale, say 10^{12} Hz, without degradation of the information. Thus, this simple optical element could have an overall digital transmission bandwidth of 10^{18} bits s^{-1} – an incredible capability. The challenge, of course, is to develop the optical elements needed to carry out logic operations on this data flow with sufficient parallelism and speed of response as to permit *some fraction* of this potential to be realized in a practical system. We shall come back to this problem shortly.

To develop prototype digital optical circuitry it is necessary to have some form of architecture in mind. If optical techniques are to compete with electronics they must offer more than large arrays of processing elements operating in parallel, and exploit, in addition, the parallel *interconnect* capacity of light. It is in this *combination* of 2d-parallelism and crosstalk-free interconnection in the remaining (3rd) dimension that optics has its advantage. Possible architectures can be classified in terms of the degree of fan-out and/or fan-in of signals between processing elements, as illustrated in Fig. 4.1.

(i) *Data transfer*: only 1-to-1 interconnect or possibly 1-to-2 fan-out is required for, say, data reordering between electronic logic processor arrays, parallel data acquisition from optical disc storage, or clock distribution with minimal time-skew.

(ii) *Cellular processing*: 1-to-4 up to about 1-to-8 fan-out could be used for applications such as high-accuracy digital image processing.

(iii) *Threshold processing*: 1-to-10 up to 1-to-100 fan-out is needed for vector-matrix multiplication or neural network concepts based in 2d cross-bar switch configurations.

(iv) *Fourier processing*: global fan-out, e.g. fan-out $> 10^6$, between an input and the Fourier plane, permits a wide range of analog optical computing concepts to be implemented.

Because, in architectures using spatially invariant interconnections, the signal fan-in levels correspond to the degree of fan-out, high accuracy *digital* processing becomes difficult for classes (iii) and (iv) above. Instead, pure analog or "fuzzy

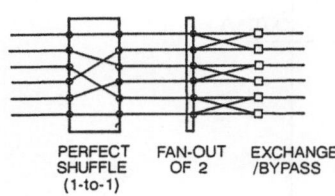

(1) FAN-OUT 1/2 e.g. REORDERING

PERFECT
SHUFFLE
(1-to-1)
FAN-OUT
OF 2
EXCHANGE
/BYPASS

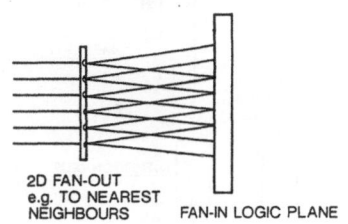

(2) FAN-OUT 4/8 e.g. CELLULAR IMAGE PROCESSING

2D FAN-OUT
e.g. TO NEAREST
NEIGHBOURS
FAN-IN LOGIC PLANE

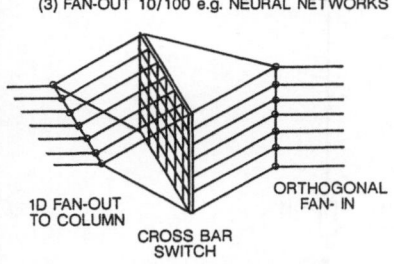

(3) FAN-OUT 10/100 e.g. NEURAL NETWORKS

1D FAN-OUT
TO COLUMN
CROSS BAR
SWITCH
ORTHOGONAL
FAN- IN

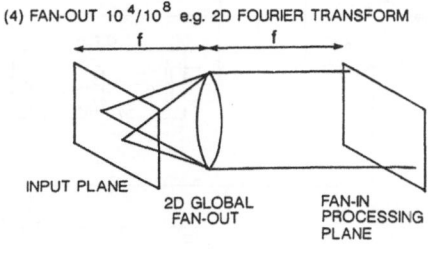

(4) FAN-OUT $10^4/10^8$ e.g. 2D FOURIER TRANSFORM

INPUT PLANE
2D GLOBAL
FAN-OUT
FAN-IN
PROCESSING
PLANE

Fig. 4.1. Schematics illustrating four different classes of optical parallel processing with increasing levels of fan-out

logic" algorithms [4.5] must be exploited. The demonstration all-optical digital circuits described later in this chapter are based on the lower fan-out approaches of classes (i) and (ii).

In the early stages of evolution of our approach to optical digital computing developed at Heriot-Watt University, it proved useful to consider, in concept, an optical processor designed to carry out a specific, albeit very simply, parallel processing task. The task chosen was to design a system, based on latchable optical logic elements, capable of performing a simple one-dimensional Ising-model calculation. Two different designs for such a processor were published by *Wherrett* [4.6] and *Walker* [4.7]. Both approaches, however, were similar and used what has since been dubbed as a "lock-and-clock" architecture in which the latching action of the logic switch arrays was used to temporarily store data before clocking it on to the next processing plane. Figure 4.2 shows a schematic for one of these processors, together with a timing diagram for the bias-power supplied to each logic array. This system shows all the main features of a loop processor of this type: input, output, programmable logic unit, memories, interconnects, power supply, clock. Processors like this are of the SIMD (Single Instruction Multiple Data-stream) type, and multiply the clock-rate of the control electronic processor by their parallelism – thus combining the advantages of optical techniques with a convenient interface to conventional electronics. It will be seen later in this chapter that some of the experimental optical digital circuits that have been subsequently constructed bear many similarities to these initial processor concepts.

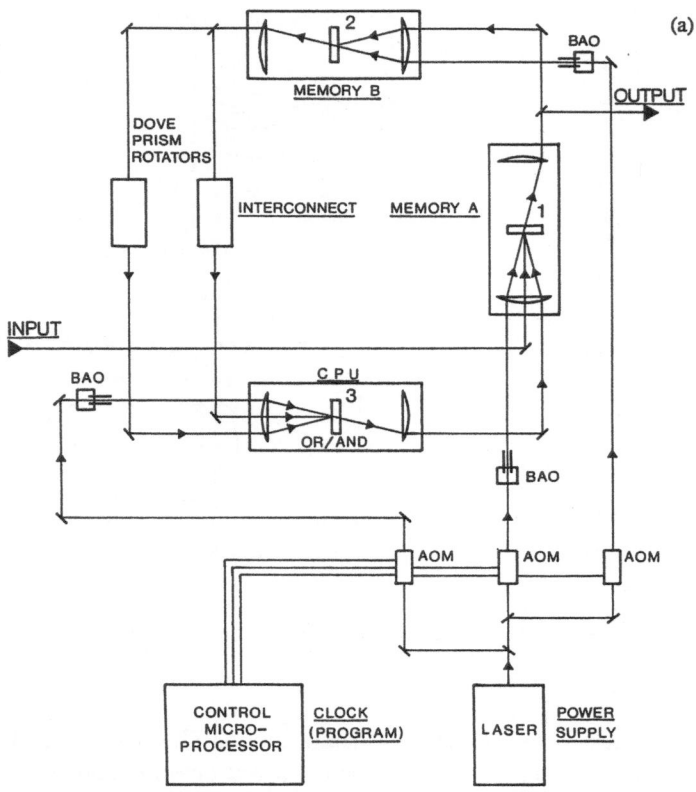

Fig. 4.2. (a) Schematic of a single-instruction multiple-data-stream optical processor. This conceptual machine, designed to carry out a simple 1d Ising model calculation, shows all the components required in an iterative processor of this type

4.1.2 Optical Logic Devices

There exist three approaches to optical logic:

(i) discrete optoelectronics, in which light is detected, the signal operated upon electronically and then converted back into an optical output by modulating or emitting a separate beam of light;

(ii) hybrid devices, in which the functions described above are integrated into a single optoelectronic structure (e.g. as in the SEED of LCLV devices); and

(iii) intrinsic nonlinear devices, in which the information remains in the optical domain and direct light–light interactions are induced by the nonlinear characteristics of the materials being used.

Approaches (ii) and (iii) can be divided further into two classes: active and passive. The active devices incorporate light-emitting sources (e.g. as in the bistable laser-diode amplifiers) and inevitably require electrical supplies. The

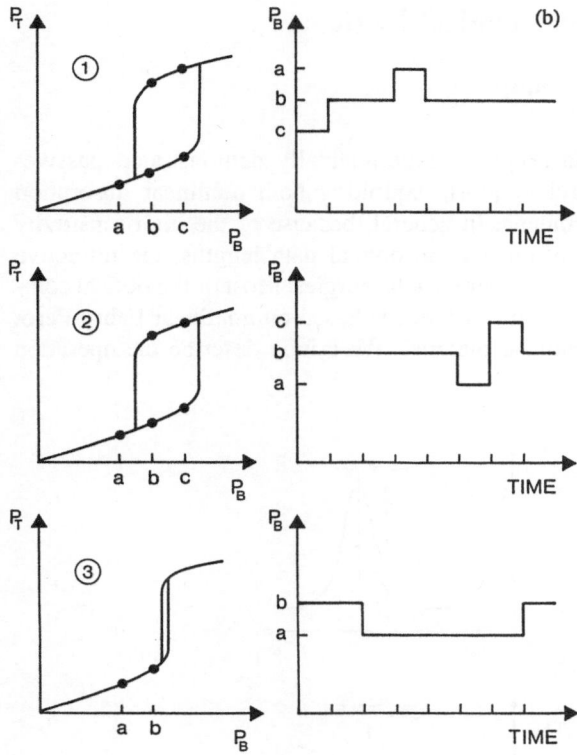

(b) *Left*: Transfer characteristics required of the nonlinear optical processing devices in system (a). *Right*: Timing diagrams for the three gate arrays, illustrating the clocking procedure

passive devices are *powered* by light inputs while the hybrid devices require electrical inputs as well.

The concept that having chosen to exploit optical interconnects one should remain in the optical domain to process the information is an attractive one, suggesting a highly efficient system that avoids multiple photon-electron-photon conversions. For this reason we concentrate in this review on the way in which passive-intrinsic optical logic devices – particularly bistable nonlinear etalons – can be used to construct optical digital processing demonstrators. The development of optically bistable devices – i.e. optical devices with two possible output states for a given set of input conditions – acted as an important spur to many of these ideas, providing a variety of binary digital components suitable for processing milliwatt-power optical signals. The next section summarizes the properties of these devices and describes the ways in which they can be used as digital logic elements.

4.2 Optical Bistability and Optical Logic

4.2.1 Refractive Optical Bistability

There currently exists a wide range of experimentally demonstrated passive-intrinsic optically bistable devices [4.8], exploiting both nonlinear absorption and nonlinear refraction phenomena. In general, because of the high sensitivity of interference devices to small changes in optical path lengths, the refractive nonlinearities lead to devices with lower switch energies. Most of the optical computing circuits described later in this chapter are based on nonlinear Fabry-Perot etalons incorporating such nonlinear material. We briefly describe the operation of this type of device.

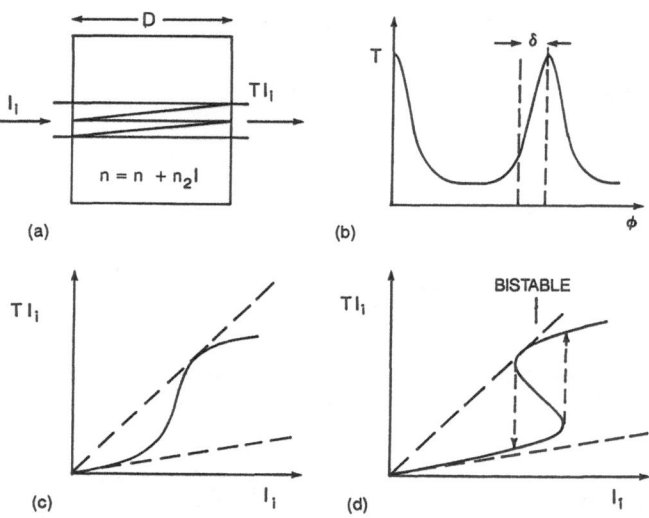

Fig. 4.3. (a) A Fabry-Perot interferometer containing a refractive nonlinear medium. (b) Variation in transmission as phase-thickness is altered. (c) A single-valued nonlinear transfer characteristic. (d) A bistable nonlinear transfer characteristic

Consider a simple Fabry-Perot cavity containing a refractive-nonlinear material as indicated in Fig. 4.3a. It is apparent that any change in internal irradiance, I, will cause a change in the relative tuning between the optical vacuum wavelength, λ, and the cavity resonances. That is, the phase thickness of the cavity φ must vary as

$$\varphi(I) = \frac{2\pi D}{\lambda}(n + n_2 I) , \tag{4.1}$$

where D is the physical cavity length, n the linear refractive index and n_2 the nonlinear refraction coefficient. The cavity transmission, T, consequently varies according to the usual Airy function description, see Fig. 4.3b,

$$T(\varphi) = \frac{T_{\max}}{1 + F \sin^2 \varphi} ,$$

(4.2)

where $T_{\max}$ refers to the maximum, on-resonance, transmission and F is the coefficient of finesse for the cavity. Thus, starting with an initial detuning from resonance, for example, of δ, as indicated in Fig. 4.3b, it follows that (assuming $n_2 > 0$) as the irradiance is increased, the cavity tunes from a low transmission towards a high transmission state. This can be summarized in the form of a transfer characteristic for the device (output/input plot) – as shown in Fig. 4.3c — which moves between the low transmission (shallow slope) and the high transmission (steeper slope) extrema. Finally, the dependence of the internal cavity irradiance I (which determines the change in refractive index) upon the resonance state of the cavity must be taken into account. For a simple cavity formed by two mirrors of reflectivity R and with negligible internal losses, this dependence can be expressed in terms of its transmission state

$$I = I_i T \frac{(1 + R)}{(1 - R)} ,$$

(4.3)

where I_i is the incident light irradiance.

Equations (4.1–3) form a cyclic set and, under the correct conditions, describe a system with positive feedback. That is, a small change in I can alter φ, and the consequent change in T will amplify the initial fluctuation of I. It is in this way that switching can occur between two bistable states. Mathematically, the solution of these equations corresponds to the plot in Fig. 4.3d, which shows the typical s-shaped curve. As the input irradiance is varied, switching occurs between the transmission states at the two turning points. The region of hysteresis,

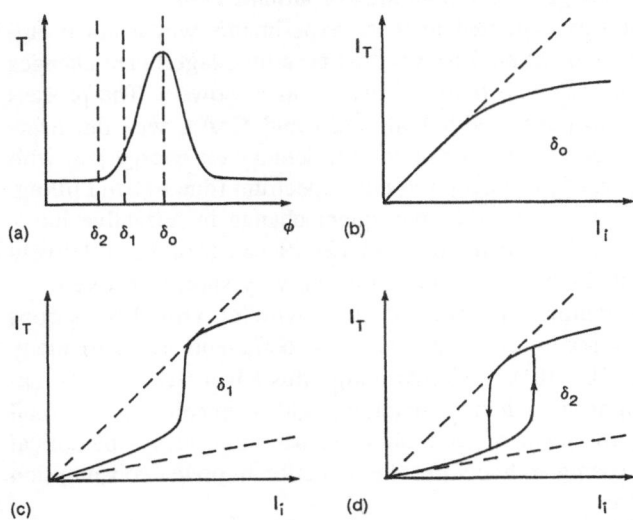

Fig. 4.4. (a) Variation in transmission of a Fabry-Perot interferometer in a region of a cavity resonance. (b–d) Various transfer characteristics for a nonlinear Fabry-Perot with initial (low signal) Φ values δ_0–δ_2

between switch-up and switch-down, is the bistable region. The third steady-state solution implied by the part of the curve with negative slope is unstable to any small perturbation in the same manner as just described for the switching action. It is therefore not observed experimentally.

An important feature of bistable devices of this type is that the transfer characteristics can be altered by changing the initial detuning between the optical vacuum wavelength and the cavity resonance. A typical sequence is shown in Fig. 4.4. If the initial detuning is zero (δ_0), then the characteristic follows a limiting response; at a critical detuning (δ_1) a single-valued vertical step can be obtained; while for larger detunings (δ_2) true bistability results.

4.2.2 Early Demonstrations of Refractive Bistability

The first experimental observations of passive intrinsic optical bistability were made at Bell Laboratories by *Gibbs* et al. [4.9] in 1976 using sodium vapor as a refractive nonlinear medium. This work utilized an atomic vapor oven contained within a Fabry-Perot resonator constructed from discrete mirrors. An important advance was the move in 1979 to exploiting nonlinearities in semiconductors demonstrated at the same time, but independently, by the Bell Laboratories group, using GaAs [4.10], and by *Miller* et al. working at Heriot-Watt University, Edinburgh, with InSb [4.11]. The experiments with GaAs used micrometer thick samples sandwiched between dielectric mirrors, while the Edinburgh group utilized the natural reflectivity of the polished surfaces of the InSb crystals to form an integrated resonant cavity of about $100\,\mu m$ overall thickness. With the incident light also focused into a $10–100\,\mu m$ wide area it became possible to contemplate the exploitation of these bistable devices as practical optical logic elements with many hundreds or thousands packed into an area of around $1\,cm^2$.

The nonlinearity that was exploited in these experiments was a newly appreciated "real-excitation" type of nonlinearity that permitted significant changes in refractive index (e.g. 0.1 %) with only milliwatt optical powers. The process underlying the nonlinear response, with both InSb and GaAs, was the near-resonant excitation of charge carriers across the semiconductor energy gap with the subsequent modification of the absorption-edge spectrum (due to band filling, exciton saturation effects, etc.), and the consequent change in refractive index as dictated by the Kramers-Kronig relations [4.12]. Because of the relatively long lifetimes of the excited carriers, compared to the very short effective excitation times in pure-$\chi^{(3)}$ nonlinear phenomena (i.e. involving virtual transitions only), these real-excitation nonlinearities can give n_2 coefficients larger by many orders of magnitude (e.g. $10^6–10^9$) [4.13]. Although this advantage is at the expense of slower time response τ, device switch energies (typically proportional to $\lambda^3 \alpha \tau / n_2$) are comparable, and when wishing to use large arrays of optical switches for parallel processing architectures the reduction in power consumption is the more important consideration.

Typical switch irradiance levels are: for GaAs, $\sim 100\,\mu W\ \mu m^{-2}$ with $\tau \sim 10\,ns$; and for InSb, $\sim 0.1\,\mu W\ \mu m^{-2}$ with $\tau \sim 500\,ns$. It is apparent from these values that an array of 10^4 optical switches of this type, each a few square wavelengths in area, could be operated by a total optical power of 10 W and would yield overall bit rates of $\geq 10^{11}$ Hz. It is extrapolations of this sort that have inspired a number of groups worldwide to investigate the possibilities of operating such optical switch arrays and the ways in which the performance of individual devices could be further improved beyond these early results. Meanwhile, a crucial question to be answered has been whether such devices could indeed be used as optical digital logic elements.

4.2.3 Optical Bistable Etalons as Logic and Memory Elements

The idea that bistable etalons could be used as digital logic devices encountered considerable scepticism. The points that have been raised encompass the importance of logic level restoration and hence gain, the need for negation (the NOT gate), the limitations of two-port devices, and the problems of threshold logic devices. Above all, those groups working on optically bistable devices were challenged to use them to construct the simplest digital circuits. It is the way this challenge was met and the approach to the construction of successfully operated circuits that is the topic of the remainder of this chapter.

(a) The Hold-and-Switch Approach to Gain and Cascadability. The first consideration is how to make a cascadable gate, i.e. a device with an output change on switching that is sufficient to induce switching of a subsequent gate, or better, one capable of driving several subsequent gates, i.e., fan-out > 1. This requires signal gain (differential gain) from the device. The method for achieving this is known as the "hold-and-switch" technique and is illustrated in Fig. 4.5. It is assumed in this description that the device is critically bistable, such that there is a sharp step in the transfer characteristic but that the region of bistability is of negligible width. This provides a nonlatching response to the input optical signals. (The same approach could equally be used, however, to obtain gain with a bistable, latching device).

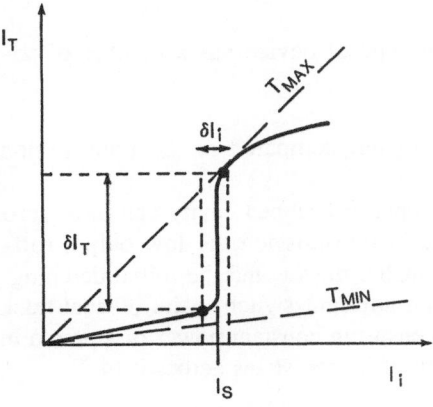

Fig. 4.5. A critically tuned nonlinear Fabry-Perot transfer characteristic illustrating the "hold-and-switch" approach to gain

As illustrated in Fig. 4.5, the transfer characteristic is bounded by two lines whose slopes correspond to the maximum and minimum transmission of the device and between which switching occurs (approximately). Typical figures for T_{Max} and T_{Min} are 70 % and 20 % respectively, implying that the magnitude of the change in output on switching, δI_T, equals 50 % of the input (in such a case). It follows that if the device is biased or "held" within, say, 10 % of switch point, I_s, an input signal $\delta I_i = 0.1\ I_s$ would give an output change $\delta I_T = 0.5 I_s$, corresponding, in this example, to a gain of 5. Clearly, more than one subsequent gate could be driven by this output.

It can be seen that the limit to the achievable gain is determined only by those technical factors that determine how closely the hold power can approach the switch irradiance, I_s, without spurious switching occurring. This is set by various factors including the stability of the laser source; the effect of mechanical vibrations; the stability of the device materials themselves; and, in arrays of switches, the reproducibility of each element and the influence of adjacent devices. These various factors and their impact on device and optical circuit design are discussed by *Tooley* [4.14].

In addition, signals that only just take the input above I_s, yield very slow switching due to the phenomenon called *critical slowing down* [4.15]. Typically, for a device held at $0.9\ I_s$ to switch in a time $< 3\tau_0$ (τ_0 being the time constant of the underlying nonlinearity), the switch irradiance must correspond to $\geq 1.3\ I_s$. This factor must also be taken into account when considering available device gain.

Under carefully controlled conditions and on short time scales only (e.g. seconds) gain factors as high as 10^4 have been demonstrated in InSb bistable devices [4.16]. More typically, gain factors of around 5 are obtained when critical slowing down is sensibly taken into account.

(b) Negation Through Reflection. The essential requirement for any binary logic circuit is the NOT operation, and this is conveniently provided by the reflection output of a nonlinear etalon device. As indicated in Fig. 4.6a, when the Fabry-Perot cavity switches from low to high transmission, its reflectivity goes in the reverse direction, from high to low, thus providing the required inverted response.

Exploiting the reflection output of this type of device has a number of advantages [4.17].

(i) The effective signal gain is always higher, compared to the transmission output.

(ii) Higher contrast can be achieved: a properly designed cavity can have zero reflectance when on-resonance, while in transmission the low output (off-resonance) level is limited by the attainable finesse and the initial detuning.

(iii) By returning the reflected signal to an adjacent switch, pairs of dual-track logic devices could be operated, thus ensuring constant power dissipation in a larger array regardless of the pattern of binary states across it [4.7].

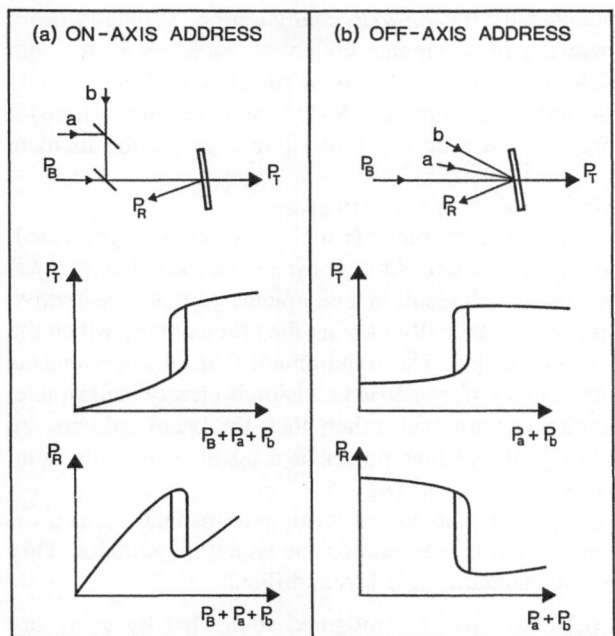

Fig. 4.6. (a) On-axis approach to addressing a bistable optical logic device, showing output powers plotted as a function of signal input. (b) As (a) but for off-axis address

(iv) Finally, excess thermal power could be dissipated in an optically opaque heat-sink attached to the rear of the array.

One further important property of the reflection mode of operation is the unique transfer characteristic that is obtained. As shown in Fig. 4.6a, the reflection characteristic passes twice through the same range of outputs on each side of the switch region, giving an alternate low-high-low output sequence as the input is steadily increased. We will discuss later how this can be exploited to make a programmable logic gate capable of providing various logic operations, including the XOR and XNOR functions.

(c) Off-Axis Address. Three Port Operation and Restoring Logic. The detailed response of bistable optical logic gates to signal inputs depends on the way in which these signals address the active area of the nonlinear etalon [4.7]. In the discussion on gain, it was assumed that the signal came along the same input axis as the hold beam, and that the two were effectively indistinguishable. This geometry has been described as "on-axis". The alternative is to bring the signals onto the device from a different direction, such that they overlap with the hold beam only within the active volume of the bistable switch. This we describe as "off-axis address".

Figure 4.6a shows schematically the *on-axis* configuration, in which beam splitters are used to combine the input signals both with each other and with the main holding beam. With this geometry, the input signals contribute directly to the transmitted and reflected outputs of the device, and the signal transfer characteristic depends simply on the total combined power of all the incident beams.

This on-axis configuration has three main drawbacks:

(i) If the beams are coherent with each other (which may often be the case), serious interference effects will occur. As a consequence, small (e.g. $\lambda/2$) changes in optical path lengths will result in a complete cycle of constructive and destructive interference, dramatically varying the efficiency by which the signals are coupled onto the device. The requirement that all interconnects between gates remain fixed to interferometric precision is clearly undesirable.

(ii) If the beams are *not* coherent with each other, then the beam splitters are inefficient, wasting both signal and bias power as a result of the minimum 3 dB loss as each beam is combined [4.18].

(iii) It can be seen from Fig. 4.6a that the output levels are not well defined on either side of the bistable region but depend on the signal magnitudes. This makes signal restoration to standard logic levels difficult.

Disadvantages (i) and (ii) above can be mitigated somewhat by using orthogonal polarizations, say for the holding beam relative to the signals. However there remains the need to combine two or more signals efficiently.

The *off-axis* configuration, as shown in Fig. 4.6b keeps the signals separate from the holding beam so that they simply induce changes in the transmission or reflectivity of the device. Only the transmitted or reflected holding beam contributes to the output, in this sense acting as a probe of the current state of the device. Thus, with a fixed holding power P_H, the transmission output P_T is constrained to lie in the range $P_H T_{Min} \leq P_T \leq P_H T_{Max}$, and similarly for reflection, $P_H R_{Min} \leq P_R \leq P_H R_{Max}$. The resulting output versus *signal*-input transfer characteristics are as sketched in Fig. 4.6b. Unlike those in Fig. 4.6a (on-axis address), these show approximately constant output levels on either side of the switch region, ensuring well-defined levels for binary logic.

Two factors must be taken into account when using off-axis address of bistable etalons. The first, the main disadvantage of this approach, is that the full 2π solid angle available on the input side of the device must be divided between the separate signal and holding beams. It follows that device dimensions corresponding to the one wavelength limit cannot be realized due to diffraction limitations on focal spot sizes. However, *Walker* [4.7] has shown that as many as 5 beams (a holding beam plus four signals) can be efficiently fanned in to a single $\sim 5\lambda$ dimension device, using a total numerical aperture of 0.5. Although the latter assumed ideal Gaussian beam focusing, conventional optical resolution criteria would still permit this level of optical (off-axis) fan-in, using, for example, 500 nm wavelength signals and a 5 μm device. In determining overall packing densities, other factors such as crosstalk and thermal dissipation are likely to be more critical parameters than this device size.

The second factor is the angular sensitivity of a Fabry-Perot cavity. As the angle of incidence changes, the initial detuning is altered and consequently not all the angularly-separated signals induce the same response from the nonlinear device. (This is a further reason why the use of a very large numerical aperture for any one input should be avoided [4.19].) This effect can either be minimized, by arranging all inputs to be at an equal angle to the etalon axis (i.e. lying on a cone about the normal [4.7]), or exploited to enhance the device response. The latter involves choosing an angle for the signals which makes them resonant with the cavity while the holding beam angle is such as to leave it detuned by the amount determined by the bistable characteristic requirements. In this way, the device will initially have maximum sensitivity to the signals, ensuring efficient switching, while, once switched, the now resonant holding beam will maintain the state of the device in the usual hysteretic manner [4.20, 21].

A nonlinear etalon with off-axis address is closely analogous to a 3-port electronic device such as a transistor, particularly when, as in the last example, it can be made more sensitive to changes in the signal input power than to comparable holding power fluctuations. Figure 4.7 illustrates this comparison. In the same way that the transistor switches its electrical conductivity according to the signal on its base, the nonlinear etalon switches its optical transmissivity according to the off-axis signal input. As long as well-stabilized power supplies are provided for both devices (either electrical or optical), then well-defined

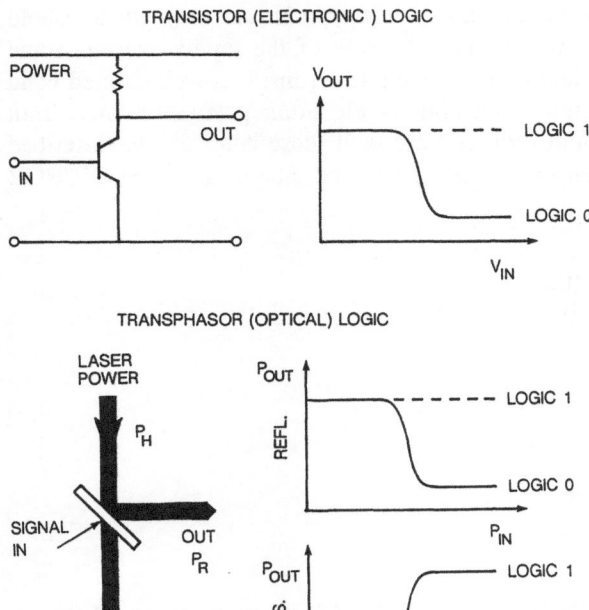

Fig. 4.7. A comparison of a transistor and an off-axis addressed nonlinear Fabry-Perot logic device (transphasor)

output logic states can be realized and extendable digital circuits with logic level restoration can be constructed.

(d) AND/OR and NAND/NOR Gates. Although, in principle, only a single logic function plus negation is required for general purpose ditigal processing, in order to minimize the total number of logic gates it is useful to have at least OR and AND gates, together with their inverted forms NOR and NAND. We shall assume that, for these gates, the off-axis configuration is the preferred option.

The ideal transfer characteristic for a logic gate shows a sharply defined step but does not need to exhibit bistability. In practice, by controlling the initial detuning from resonance, a narrow bistable region can be obtained which gives clear switching, while, when biased at an input level just below this region, permitting the gate to self-reset once the input signals are removed. Figure 4.8 shows how a two-input transmissive gate can act as either an OR or an AND gate. The changeover between these two responses can be achieved by changing the holding power so as to adjust the proximity of the bias power to the switch power by an amount equivalent to a single logic-1 input. Provided that the signal powers are small in comparison to the holding beampower (which is essential if significant gain is to be obtained), then the difference in output logic levels due to this change in "power-supply" will be tolerable. Alternatively, the device could be operated as a 3-input gate with constant bias, the extra input determining the required function [4.7, 22]. The inverted NOR and NAND functions are, of course, obtained by utilizing the reflection output.

One of the criticisms of logic gates of this type is that they are threshold devices, i.e. devices which switch when the *sum* of the inputs exceeds some threshold. This requires the input signal levels to lie inside a well-defined band with both upper and lower limits. Conventional electronic gates set a lower limit on a logic-1 but no upper limit (within reason). If these latter can be described as nonthreshold devices, then the off-axis addressed, nonlinear etalon OR/NOR

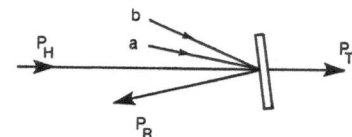

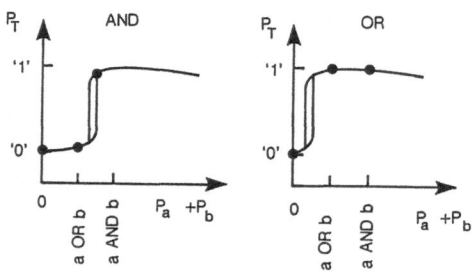

Fig. 4.8. Illustration for an off-axis addressed optical logic device, of how a change in holding power can give AND or OR logic functions

gate is not a threshold device either. It simply has a minimum recognizable input which will ensure switching and a region with roughly constant output, above the switch point, which is, over a reasonable range, insensitive to the magnitude of the input. The optical AND/NAND gate, as described, *is* a threshold device, however, and has the disadvantage that one of the two inputs could alone cause switching if it were large enough. This need not be a serious problem provided the AND gate is limited to a 2-input form only. Alternatively, other AND-gate configurations could be considered [4.7] or one could revert to using NOR-gates only, in the appropriate combinations.

(e) Memory Elements. An essential feature of any processor is a short-term memory from which the logic unit can read and to which it can write. Bistable nonlinear etalon devices, biased into their bistable region, act as latching switches – continuously emitting a signal corresponding to their current state. They therefore provide the required memory function.

Bistable memory elements can be used within an optical digital parallel processing loop in two ways:

(i) They can be clocked between a "read/write" state, just below the switch power, and a "reset" state, with zero or near-zero input power, as proposed by *Wherrett* [4.6].

(ii) They can be clocked between three separate hold points: "write", just below switch point; "read", further below switch point such as to be immune to changing input signals but still in the bistable region (and hence storing data); and "re-set", anywhere below the region of bistability.

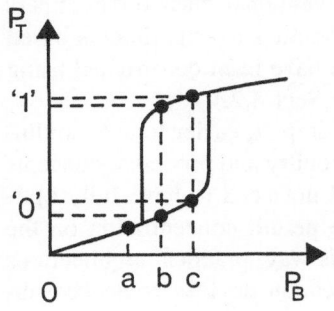

Fig. 4.9. A bistable transfer characteristic suitable for a memory device. Bias at *a*: reset; at *b*: store only (read); at *c*: write

This latter approach, illustrated in Fig. 4.9, was proposed by *Walker* [4.7] as a method of cycling arrays of 2d digital information around a processing loop using only two bistable memory arrays. The alternative method, described above, uses three such arrays to ensure the data is correctly buffered within the loop, but does have the advantage of a simpler (and less critical) clocking cycle. In both instances the memory is temporary, or volatile, in that the removal of the bias beam (array) permits the 2d spatial refractive index modulation to relax and thereby resets the device.

4.2.4 Practical Aspects of Devices for Experimental Optical Digital Processors

Although the early work on optically bistable InSb or GaAs etalons (described in Sect. 4.2) gave great encouragement to the idea of constructing optical digital processors, both devices had a number of drawbacks which made them inconvenient for demonstration processor construction. For example, the major limitation of GaAs bistable devices is the poor ratio of optoelectronic to optothermal nonlinear response. That is, as a consequence of a lower n_2 coefficient at the shorter wavelength (n_2 typically scales as the inverse cube of the semiconductor energy gap), the device must be operated in very close resonance with the absorption edge and consequently the dissipated power induces significant temperature rises and unwanted thermally induced index changes. This heating effect has prevented continuous, steady-state operation of GaAs bistable etalons, greatly limiting the experimental circuits that might be constructed. This problem may be solved in the future by using material with longer carrier lifetimes and hence a larger n_2 (as discussed in Sect. 4.2.2) or by improved heat-sinking, particularly in conjunction with fully integrated cavity structures [4.23, 24]. InSb bistable etalons, on the other hand, work very well in a steady-state mode but instead have two technical limitations. Firstly, to tune the band gap into near resonance with a suitable high power cw laser (CO laser), they must be cooled to liquid nitrogen temperatures. Although this may not be a fundamental problem – any powerful all-optical processor will need to dissipate a lot of heat, and liquid-nitrogen would be an excellent coolant – it is not very convenient when attempting to construct prototype circuitry. A further inconvenient aspect is the 5.5 μm working wavelength. This lies outside the range of conventional glass optics, fibers and holographic elements and once more adds difficulties to experiments based around these devices. (Nonetheless, simple circuits have been constructed using coupled InSb switches and are described further in Sect. 4.2.)

From the above discussion, we concluded that simpler, easier to use, nonlinear etalon devices were required if optical digital circuitry and processor concepts were to progress experimentally. Such devices did not need to have fully optimized performance but had to be good enough to permit concentration on the circuit concepts rather than device problems. In this way, practical architectures could be developed and other improved bistable etalon devices could be substituted as they became available. It was on the basis of this approach that we developed the nonlinear interference filter devices described in the next section.

4.3 Nonlinear Interference Filters as Optical Logic Elements

4.3.1 Bistability in Interference Filters

As a consequence of optically induced heating, a multilayer thin-film interference filter with some internal absorption shows a nonlinear response, with respect to increasing incident irradiance, if it includes materials with significant thermo-

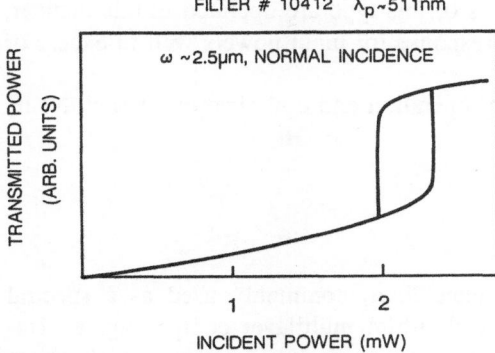

FILTER # 10412 $\lambda_p \sim 511$nm

$\omega \sim 2.5\mu$m, NORMAL INCIDENCE

TRANSMITTED POWER (ARB. UNITS)

INCIDENT POWER (mW)

Fig. 4.10. An example of a bistable transfer characteristic for a ZnSe NLIF operating at 514 nm (5 μm diameter spot). Switch-up power: 2.3 mW; switch response time: $\sim 50\,\mu$s

optic (dn/dT) coefficients. The standard interference filter design corresponds to a thin-film version of the Fabry-Perot cavity and therefore, as with the nonlinear etalon devices discussed earlier, optically bistable switching can be induced. These optothermal devices are particularly convenient for a number of reasons. They work within a range of visible and near-infrared wavelengths, at room temperature, in a continuous steady-state fashion, and with switch powers down to about 1 mW. In addition, they rely on well-established optical coating techniques, which can be exploited to fabricate large area devices with good spatial uniformity. Finally, they can be made from a range of materials and deposited on a variety of substrates, hence permitting convenient optimization and thermal engineering to suit specific circumstances and applications.

Figure 4.10 shows an example of a transmission transfer characteristic for a ZnSe nonlinear interference filter (NLIF) operating with a 514 nm wavelength input. Figure 4.11 shows a similar device, with an off-axis signal input, operating in both the self-resetting and latching modes. It can be seen that clearly defined

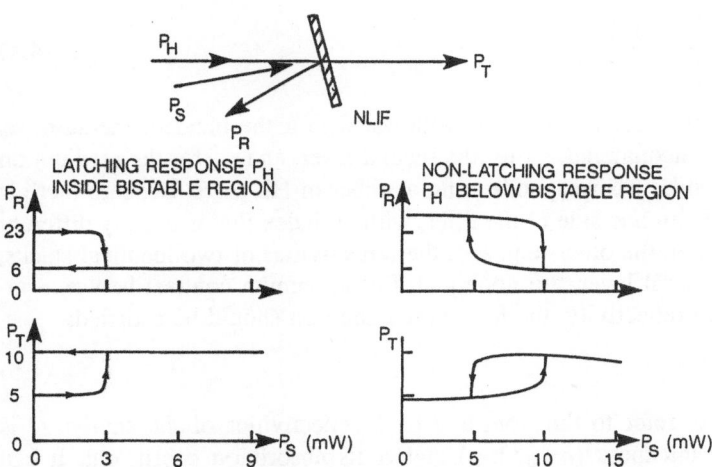

Fig. 4.11. Examples of transmission and reflection responses of an off-axis addressed ZnSe NLIF. *Left*: held in the bistable regime, $P_H = 35$ mW. *Right*: held at a power below the bistable region, $P_H = 20$ mW. 514 nm operation, spot diameter $\sim 80\,\mu$m

107

output logic levels exist when devices of this type are operated in this manner, and that there is a strongly limiting response for input powers well in excess of the switch threshold.

This section describes the design, operation and optimization of such NLIF optical logic devices.

4.3.2 Linear Interference Filters

The basic narrow-band-pass interference filter, commonly used as a spectral (spike) filter, is typically a dielectric thin-film multilayer comprising: a "1/4-wave" reflecting stack, deposited on the substrate, an $m\lambda/2$ optically thick spacer layer placed on top of this reflector (m being an integer), and finally, a second reflecting stack to complete the cavity. The overall structure can be summarized in the following manner:

Substrate $\cdot$ j(HL) $\cdot$ m(HH) $\cdot k$(LH) $\cdot$ air ,

where L and H refer to $\lambda/4$ optical thickness low and high index layers, respectively. The integers j and k determine the reflectivity of the stacks – the more HL pairs, the higher the reflectivity (within limits set by internal absorption) – and the integer m determines the order of the cavity. The nonlinear interference filters commonly employ $j, k = 2\text{–}4$ and $m = 1\text{–}8$. The low index materials are typically ThF$_4$ or cryolite and high index materials, ZnSe or ZnS. More complex multi-cavity structures can also be constructed, with two or three spacer layers sandwiched between reflecting stacks, to provide stronger suppression of the wings of the bandpass peak.

The stack reflectivities can be calculated using the relation

$$R_j = \left(\frac{n_a n_c^{2j} - n_d n_b^{2j}}{n_a n_c^{2j} + n_d n_b^{2j}} \right)^2 , \tag{4.4}$$

where the refractive indices, n, are as follows: n_a for the incident medium, n_b for the first layer encountered, n_c for the second layer, and n_d for the medium on the transmission side. j corresponds to the number of HL pairs. Because there is always a substrate on one side of the filter, with an index that is usually different from the medium on the other (e.g. air), the reflectivities of two identical stacks, surrounding a spectral layer, are not equal. For maximum contrast between on- and off-resonance reflectivity, the following condition should be satisfied:

$$R_{\text{F}} = R_{\text{B}} \, e^{-2\alpha D} , \tag{4.5}$$

where R_{F} and R_{B} refer to the front and back reflectivities of the cavity; D is the spacer layer thickness ($m\lambda_0/2n_{\text{H}}$) and α its absorption coefficient. It can be seen that if absorption losses are negligible ($\alpha D \ll 1$) extra layers must be incorporated into one of the stacks to ensure the condition $R_{\text{F}} \sim R_{\text{B}}$. Similarly,

R_F and R_B must be adjusted correctly for the case where internal cavity losses are significant ($\alpha D \geq 1$) according to (4.5).

To a first approximation, a multilayer dielectric interference filter can be treated as equivalent to a solid etalon, of thickness equal to the spacer layer, having mirrored surfaces with reflection coefficients R_F and R_B. However, a more complete description includes the effect of the variable phase changes that can be induced on reflection from the stacks. Only at normal incidence and at the design wavelength (i.e. corresponding to H and L being exact $\lambda/4$ optical thicknesses) is the phase change on reflection from the stacks zero. *Pidgeon* and *Smith* [4.25] have shown that the phase change on reflection from a 1/4-wave stack, $\Delta\varphi$, varies with wavelength λ (for normal incidence) as

$$\tan \Delta\varphi \propto \sin(\pi\lambda_0/\lambda) , \tag{4.6}$$

and with external angle of incidence θ (for $\lambda = \lambda_0$) as

$$\tan \Delta\varphi \propto \sin^2 \theta . \tag{4.7}$$

Because angular adjustment of the cavity is a convenient method of tuning the resonance point, and because various wavelength detunings are employed when using interference filters as nonlinear devices, it is often important to include phase-change effects.

For a simple Fabry-Perot etalon, with constant phase-change reflectors, the dependence of peak wavelength λ_p upon the internal angle of propagation θ_c, relative to the cavity axis, is

$$\lambda_p(\theta_c) = 2nD \cos \theta_c . \tag{4.8}$$

This can be rewritten in terms of the external angle of incidence θ,

$$\lambda_p(\theta) = \lambda_0 \left(1 - \frac{\sin^2 \theta}{2n^2} \right) , \tag{4.9}$$

where $\lambda_0 = \lambda_p(0)$. The shift in peak wavelength, $\Delta\lambda_p(\theta)$, for a cavity with multilayer dielectric reflecting stacks can be deduced from a corresponding equation in which the etalon index, n, is replaced by an effective index n_e that takes into account the different apparent optical thickness of the cavity resulting from nonzero phase changes on reflection. Thus for a thin-film interference filter:

$$\Delta\lambda_p(\theta) = \frac{\lambda_0 \sin^2 \theta}{2n_e^2} , \qquad \text{where} \tag{4.10}$$

$$n_e^2 = \frac{mn_s(n_H - n_L) + n_L n_H}{(m/n_s)(n_H - n_L) + n_L/n_H - 1} . \tag{4.11}$$

In the above, n_s refers to the spacer index, to include those cases in which it is different from the high index stack material. It can be seen that for high-order cavities, when $m \gg 1$ (e.g. bulk etalons with multilayer dielectric reflectors) the effective index reduces to n_s, as expected.

One consequence of the dependence of λ_p upon θ is that when light is focused to a very small spot-size onto a nonlinear interference filter (e.g. to maximize the irradiance) a significant reduction in the effective finesse can be induced as a consequence of the larger range of angles contained within the cone of incident radiation. This leads to an optimum spot size below which the switch power increases despite the higher irradiance levels [4.19].

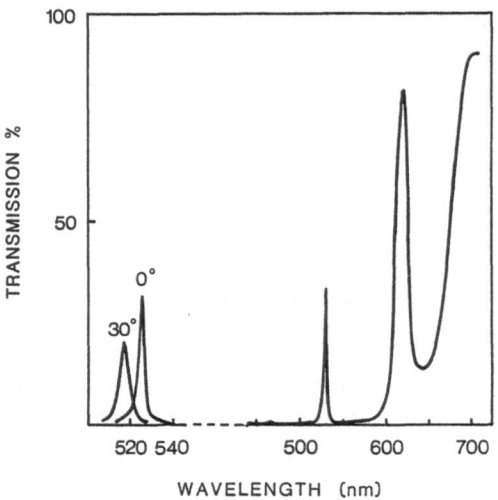

Fig. 4.12. *Right*: Transmission spectrum for a typical ZnSe NLIF. *Left*: Shift of the narrow-band-pass peak with changing angle of incidence

Figure 4.12 shows the linear spectrum of a typical filter, of the type being used to make optically bistable devices for the 514 nm argon-laser wavelength. The spacer consisted of $\sim 400\,\text{nm}$ of ZnSe and the stacks were constructed from ZnSe (H) and ThF$_4$ (L). Using $j = k = 3$, the two stack reflectivities were of order 90 %–95 %. The spectrum consists of a 4 nm wide transmission peak at 524 nm, separated by a large region of high reflectivity from the next peak at $\sim 620\,\text{nm}$. The high transmission at longer wavelengths occurs because the stack reflectivities have diminished; and, at wavelengths shorter than $\sim 500\,\text{nm}$, ZnSe absorption dominates and both transmission and reflection are low. The figure also shows how angle tuning can be used to shift the peak to the required detuning relative to the laser wavelength.

4.3.3 Nonlinear Interference Filters – Early Experiments

Passive-intrinsic optical bistability was observed for the first time within a semi-conductor by *Karpushko* and *Sinitsyn* [4.26] in 1978, using a nonlinear interference filter (NLIF) with a ZnS spacer. Although at the time the mechanism underlying the nonlinearity was not understood, subsequent studies by *Weinberger* et al. [4.27] and *Smith* et al. [4.28] have demonstrated its optothermal nature. The paper by *Smith* et al. [4.28] was the first report of optical bistability in a NLIF with a ZnSe spacer and with reproducible switching characteristics. Early work had shown highly irreproducible device responses as a result of effects such as desorption of water from porous dielectric layers [4.27]. Slow switch power drift has continued to be a problem with some of these devices [4.29] but is now being overcome by using structurally more stable thin-film material (e.g. deposited by UHV molecular beam techniques) and different operating wavelengths [4.19, 29, 30].

In 1984, *Olbright* et al. [4.31] described the use of small focal spot-sizes ($\sim 10\,\mu$m) to obtain switching in both ZnS and ZnSe NLIFs on a time scale of $\sim 20\,\mu$s. At around the same time, *Apanasevich* et al. [4.32] showed that by using high thermal conductivity (crystalline) substrates, time responses of less than 1 μs could be achieved.

Results of this type, together with switch powers in the region of 10 mW, encouraged more detailed studies of the way in which NLIFs operate and the extent to which they might be optimized.

4.3.4 Theory and Optimization of NLIFs

The parameters determining the change in refractive index within a NLIF spacer, and hence its tuning, are the thermo-optic coefficient of the spacer material and the temperature rise induced by the incident light. At equilibrium, the change in temperature is determined by a balance between the absorbed incident optical power and the cooling rate. The latter is characterized by a thermal time constant, determined by the manner in which the sample is heat-sunk. In most experiments with NLIF devices, the substrate acts as a heat-sink located just 1-2 μm from the volume within which the heat is being generated – the spacer layer. Because typical laser spot sizes are $> 5\,\mu$m diameter, the heat flow is dominantly normal to the NLIF layers, and the NLIF can be approximated to an infinitely thin heat source on the substrate surface. On the basis of this geometry, it can be shown [4.33] that the thermal time constant τ is given by

$$\tau \sim \frac{r_0^2 C_p \varrho}{4\kappa_s} \,, \tag{4.12}$$

where r_0 is the spot radius and $\kappa_s / C_p \varrho$ is the thermal diffusivity of the substrate. The same equation can be obtained by assuming a purely one-dimensional heat flow to a perfect heat-sink at a distance $r_0/2$ below the substrate surface [4.34]. It can therefore be concluded that the effective volume V that is being heated is $\sim \pi r_0^3/2$. Given this volume and the requirement that heating and cooling rates must balance, the equilibrium temperature rise for an incident power P is

$$\Delta T = \frac{AP\tau}{C_p \varrho V} = \frac{AP}{2\pi \kappa_s r_0} \,, \tag{4.13}$$

where A is the fraction of power absorbed, and κ_s is the substrate thermal conductivity.

An estimate of the power required for bistability can be made by noting that the minimum change in spacer optical thickness necessary to induce a switch, assuming a finesse $\mathcal{F}$ of more than ~ 10, is $\sim \lambda/(\mathcal{F}2\sqrt{3})$. This corresponds to a temperature rise

$$\Delta T_s = \frac{\lambda}{2\sqrt{3}\mathcal{F}D(dn/dT)} , \tag{4.14}$$

where dn/dT is the thermo-optic coefficient. In the same high-finesse limit the absorptance A_s at the critical switch power is

$$A_s = \frac{3(1 - R_F)(1 + R_B e^{-\alpha D})(1 - e^{-\alpha D})}{4(1 - R_\alpha)^2} , \tag{4.15}$$

where

$$R_\alpha = \sqrt{R_F R_B}\, e^{-\alpha D} . \tag{4.16}$$

(N.B. For a high-finesse, symmetric cavity, optimized in the manner discussed at the end of this section, $A_s \sim 1/3$.)

Assuming typical values for a ZnSe NLIF: wavelength $\lambda = 0.5\,\mu m$, $\mathcal{F} = 20(R_F = R_B \sim 0.95, \alpha D \sim 0.01)$, $n \sim 2.7$, $D = 0.75\,\mu m$ ($m = 8$) and $dn/dT \sim 2 \times 10^{-4}\,K^{-1}$; then $\Delta T_s \sim 50\,K$. From (4.13) it can then be concluded that, assuming a glass substrate with $\kappa_s \sim 0.01\,W\ cm^{-1}$ and using $A_s \sim 0.3$, the switch power per spot diameter ($2r_0$), given by

$$\frac{P_s}{2r_0} = \frac{\Delta T_s \pi \kappa_s}{A_s} , \tag{4.17}$$

should be of the order of $\sim 0.5\,mW\ \mu m^{-1}$. This is consistent with typical ZnSe NLIF experimental results: e.g. $P_s \sim 50\,mW$ for $2r_0 \sim 100\,\mu m$ [4.35].

The time constant can be estimated similarly from (4.12) using $C_p\varrho = 2\,J\,K^{-1}$ for glass, to obtain $\tau/(2r_0)^2 \sim 10^{-7}s\ \mu m^{-2}$. This is again consistent with experiment, e.g. $\tau \sim 1\,ms$ for $2r_0 \sim 100\,\mu m$ [4.33].

The proportionalities of switch power, $P_s \propto r_0$, and time constant, $\tau \propto r_0^2$, were demonstrated experimentally by *Janossy* et al. [4.33, 35] and these results are reproduced in Fig. 4.13.

Wherrett et al. [4.36] have carried out more detailed calculations of bistable NLIF performance, including the phase effects resulting from light-induced index changes in the reflecting stacks. They conclude that in most circumstances the switching characteristics of a NLIF device can be analyzed on the basis of linear reflectors. Assuming again a large illumination spot, relative to the film thickness, the critical switch power can be written

$$P_s = \frac{\lambda \alpha \kappa_s r_0}{2(dn/dT)} \left(\frac{f(R_F, R_B, \alpha D)}{\alpha D} \right) . \tag{4.18}$$

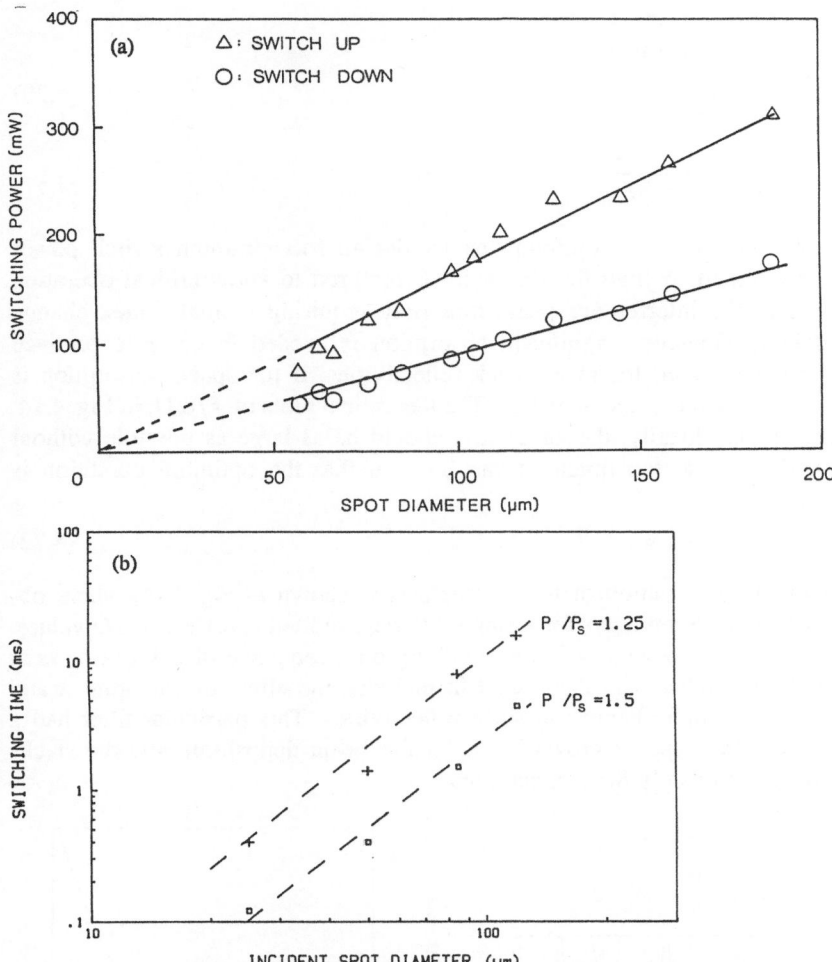

Fig. 4.13. (a) An example of ZnSe NLIF switch power variation with illumination spot diameter, showing a linear dependence. (b) An example of ZnSe NLIF switch-time variation with illumination spot diameter. The slope indicates a proportionality to the spot diameter squared, and the dependence on P/P_s (the degree of overswitch) is a manifestation of the phenomenon "critical slowing down"

This expression is a more complete version of that used above for estimating high finesse cavity switch powers (4.14, 17). The difference is that the absorptance A (which is a function of the cavity parameters and its tuning), and the critical detuning (determined by the cavity finesse) have been incorporated into the term $\alpha[f(R_F, R_B, \alpha D)/\alpha D]$. The function f can be written, with no further approximations, as

$$f = \frac{(1 - R_\alpha)^2}{(1 - R_F)(1 + R_B e^{-\alpha D})(1 - e^{-\alpha D})} \frac{\sqrt{2}}{16} \frac{[3(F + 2) - d]^2}{[(F + 2)d - (F + 2)^2 - F^2]^{1/2}} \,, \tag{4.19}$$

where

$$F = \frac{4R_\alpha}{(1 - R_\alpha)^2}$$ (4.20)

and

$$d = \sqrt{(3F + 2)^2 - 8F} .$$ (4.21)

Using these relations, the optimum cavity design for minimum switch power can be determined. A high finesse cavity is required to allow critical operation very close to the interference peak, thus only requiring a small index change for switching. However, significant absorption is needed in order to produce the index change and, for given stack reflectivities, if the spacer absorption is increased too much the finesse is lost. The theoretical plots of $f/\alpha D$, in Fig. 4.14, demonstrate that, ideally, the absorption should be as large as possible without reducing the finesse too much. It can be seen that the optimum condition is approximately

$$\alpha D = 2 - R_F - R_B .$$ (4.22)

Experimental confirmation of these principles is shown in Fig. 4.15, where observed critical switch powers for a single filter are plotted at different αD values. In this experiment, α was varied by working on a sequence of cavity orders at different wavelengths. The theoretical fit included the effect of changing λ and the consequent small changes in stack reflectivities. This particular filter had a 4.3 μm thick ZnSe spacer grown by molecular beam deposition, and the stacks provided approximately 80 % reflectivity.

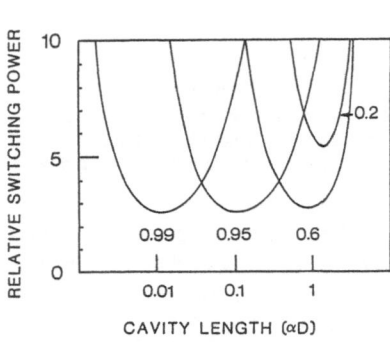

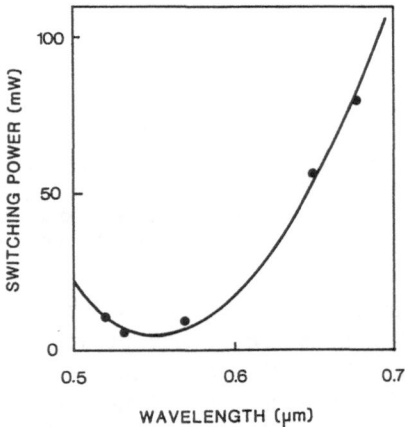

Fig. 4.14. The function $f/\alpha D$ (proportional to switch power, assuming a constant absorption coefficient α) as a function of the length of the cavity D (scaled by α). Each curve corresponds to different values of (equal) front and back reflectivities, as indicated

Fig. 4.15. Critical switch power versus wavelength for a ZnSe NLIF with $\sim$ 80 % reflecting stacks and a 4.3 μm-thick spacer (grown by molecular beam deposition). Dots are experiment, and line is theory. The dominant effect of increasing the wavelength is a reduction in absorption coefficient away from the band edge

4.3.5 Performance Limits of NLIFs

As has been discussed in the previous section, the critical switch power for a bistable NLIF and its thermal time constant obey the following proportionalities (see also Fig. 4.13):

$$P_s \propto \kappa_s r_0 \tag{4.23}$$

$$\tau \propto \frac{r_0^2}{\kappa_s} . \tag{4.24}$$

It can be seen from this that the use of smaller spot radii r_0 both reduces the required power and improves the speed of response. On the other hand, the substrate conductivity κ_s can be used to control a trade-off between faster operation and greater sensitivity. The product $P_s\tau$ corresponds to a characteristic switch energy E_s, which is independent of κ_s but proportional to r_0^3:

$$E_s = P_s\tau \sim \frac{\lambda C_p \varrho \pi r_0^3}{4\sqrt{3} A_s \mathcal{F} D(dn/dT)} . \tag{4.25}$$

The factor πr_0^3 reflects the dependence of the switch energy E_s on the volume of material being heated. This dependence provides a strong motivation to go to small spot sizes. However, two factors limit the expected advantage as the spot diameter is reduced below $\sim 10\,\mu m$. Firstly, the effective cavity finesse is reduced, as noted in Sect. 4.3.2, due to the increase in the incident light cone angle; and secondly, the heat-flow geometry changes.

For small r_0, (4.25) breaks down because the extent of the transverse thermal gradients becomes comparable to (or greater than) the spot size. Longitudinal conduction is then no longer dominant, and the effective volume is determined not by the illumination spot size but by the distribution of heat diverging from an effective point source. Thus, we observe for a spot radius of $2\,\mu m$ on an $m = 8$ ZnSe NLIF, $P_s \sim 2\,\text{mW}$ and $\tau \sim 50\,\mu s$ (switch-off time), giving a measured $E_s = 100\,\text{nJ}$, compared to a calculated switch energy (4.25) of $E_s = 2\,\text{nJ}$. To bring the experimental result more into line with this theoretical value, it may be necessary to isolate the illuminated areas by etching thermal-barrier grooves around each active element, as shown in Fig. 4.16. Arrays of such "pixellated" NLIF elements, besides having lower switch energies, will also, as a result of the inhibited transverse conduction, have much reduced crosstalk between pixels [4.37]. They can also act as microresonators [4.38], overcoming the optical spot-size limitation by guiding the light, with low angular spread, through the interference structure.

Finally, in considering micrometer-dimension NLIF pixels, account must be taken of the finite thickness of the thin-film multilayer. A significant temperature gradient can exist within this region, and consequently the thermal properties will no longer be determined by the substrate but by the thin-film materials themselves. (Indeed, the glass substrate could then be replaced by a high conductivity material, such as sapphire, to aid the removal of heat from the array without

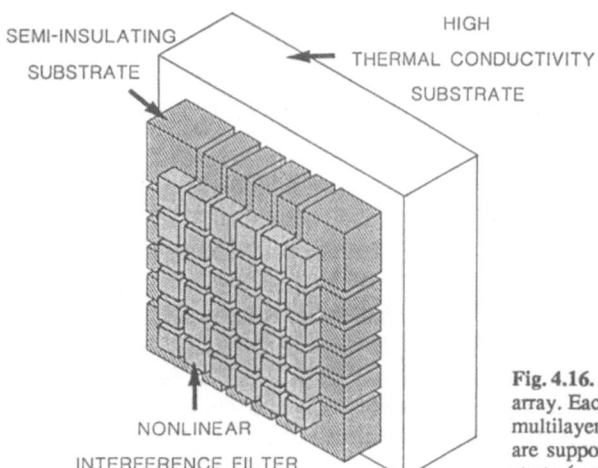

SEMI-INSULATING SUBSTRATE

HIGH THERMAL CONDUCTIVITY SUBSTRATE

NONLINEAR INTERFERENCE FILTER

Fig. 4.16. An example of a pixellated NLIF array. Each pixel corresponds to a thin-film multilayer NLIF on a thin insulator. They are supported on a high conductivity substrate (e.g. sapphire)

affecting the performance of individual devices.) The ideal situation is one in which the longitudinal temperature gradient of a transversely isolated pixel is contained entirely within the spacer layer. This ensures that only the active region of the device is significantly heated, giving maximum efficiency. Under such a condition, (4.25) can be rewritten by assuming that the effective volume πr_0^3 is replaced by $\pi r_0^2 D$ to obtain

$$\frac{P_s \tau}{\pi r_0^2} = \frac{\lambda C_p \varrho}{2\sqrt{3} A_s \mathcal{F}(dn/dT)} . \tag{4.26}$$

Using the same values as used with (4.14) and (4.17), except for changing $C_p \varrho$ from the value appropriate for glass to that for the ZnSe spacer ($1.8\,\mathrm{JK^{-1}\,cm^{-3}}$), a specific switch energy of 190 pJ $\mu\mathrm{m}^{-2}$ is obtained.

4.3.6 Bistable Etalons with Absorbed Transmission

An alternative strategy for NLIF cavity optimization is to introduce absorption in regions of the structure other than in the spacer. Thus, for example, thin metallic coatings around a nonabsorbing spacer can be employed, where the absorption in the metal is used to heat the spacer by thermal conduction [4.36]. The ideal arrangement is to place such an absorbing layer (metal, semiconductor or other high absorption coefficient material) on the output face of a high finesse cavity [4.39]. This structure, known as a bistable etalon with absorbed transmission (BEAT), can be simultaneously optimized for high finesse *and* high absorption – particularly if the device is used in reflection, when the absorber can be made fully opaque.

Many of the circuits described in the following section are based on NLIF/ BEAT devices. They have proven particularly useful for digital optics experiments for a number of reasons. (i) They represent a good 3-port device with the absorber layer providing an input port spatially separated from the power-input (holding beam) which is incident on the opposite face of the interference filter. This is equivalent, in response terms, to the off-axis geometry described in Sect. 4.2.3. (ii) Lower switch powers can be realized and, because the signal is efficiently absorbed, higher gain can be achieved. (iii) The signal can be incident at any angle, avoiding the problem of coupling into the angle-sensitive interference structure [4.21]. (iv) Broad-band absorbers such as metals or narrow-gap semiconductors can be used over a wide range of wavelengths and only the position of the interference peak needs to be adjusted, by scaling the thickness of the layers, to match a different holding beam wavelength (dn/dT does not vary rapidly with wavelength).

The latter advantage has been exploited to make ZnSe bistable switches for use with 830 nm diode lasers [4.19]. This is an important advance as the high power GaAlAs-diode near-infrared lasers, now becoming available, are particularly convenient, highly efficient, compact sources. A further advantage of working at this wavelength is that the rate at which the switch power drifts during operation is much reduced, compared to a similar device operating at 514 nm. This lower drift rate is thought to result from reduced photostructural effects at the lower photon energy [4.19]. An example of a bistable characteristic from an 830 nm ZnSe NLIF/BEAT device is shown in Fig. 4.17.

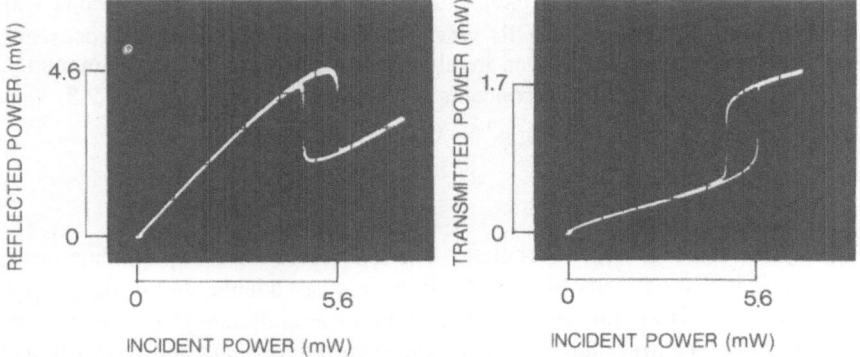

Fig. 4.17. Reflection and transmission transfer characteristics for a ZnSe NLIF/BEAT device. The partial transmission is a result of using, in this example, a 70 % absorbing aluminium film (< 10 nm thick) as the opaque layer. These results were obtained using an 836 nm diode-laser source (spot diameter 15 μm)

The same parameters as assumed for ZnSe NLIF devices in the previous section can be taken to predict the optimized performance of a BEAT device. Again (4.26) can be used, but in this case it is assumed that (i) $\alpha = 0$ – hence (for $R = 0.95$) the finesse increases to 60 – and (ii) the absorber layer is fully opaque – making the absorptance at the critical switch power $A_s = 3/4$. This reduces the specific switch energy to ~ 40 pJ μm^{-2}.

Assuming that a higher finesse can be realized, e.g. $\mathcal{F} \sim 150$ (reflectivities of 0.98), then a 1 μm square pixel would have a switch energy of 15 pJ. This implies that a processing capability of over 5×10^{10} gate-operations per second could be attainable in, for example, an array of 10^4 pixels powered by 1 W of optical input.

4.4 All-Optical Digital Computing Circuits

4.4.1 Background

Despite considerable worldwide research into nonlinear optical devices with claimed potential as logic elements, until around 1985, relatively few experiments involving their assembly into working circuits had been carried out. As already noted in Sect. 4.1, this was used by sceptics as evidence that optical digital systems were impossible to construct [4.40]. Motivated by a wish to respond to these criticisms and a desire to develop practical nonlinear devices for real optical processing systems, our group at Heriot-Watt University, Edinburgh, and the group headed by Hyatt Gibbs at the University of Arizona, Tucson, set about constructing a series of all-optical digital circuit demonstrations and processing experiments.

These experiments, reviewed in the remainder of this section, were mainly based around the NLIF bistable devices described in Sect. 4.3. The circuits and systems constructed were generally very simple with little or no processing power. However, they have proven invaluable in demonstrating, experimentally, the validity of the optical logic concepts outlined previously in Sect. 4.2.3.

4.4.2 Coupled-Switch Circuits

(a) **Coupled InSb Device Circuits.** The earliest circuit experiments were demonstrations that optically bistable devices were cascadable, that is, the change in output obtained from one device as it switched was sufficient to initiate switching of a second. The first such demonstration, reported by *Smith* and *Tooley* [4.41] at the 1983 Optical Bistability Conference, utilized two InSb bistable elements. Each was separately biased, the transmitted or reflected output of one element being directed at the other. As a consequence of the optical arrangement employed and self-defocusing effects in the first element, the transmitted irradiance reaching the second element, in practice *fell* on switching, and hence it was shown that switch *off* could be initiated.

Walker et al. [4.42, 43, 20] constructed the first such circuit to make efficient use of the light output from the first element. In this case, the interconnect between the switches was a simple imaging system with sufficient aperture to ensure that defocusing effects were negligible. The second element thus correctly responded to the total change in power output from the first element. This experiment was interesting for two further reasons:

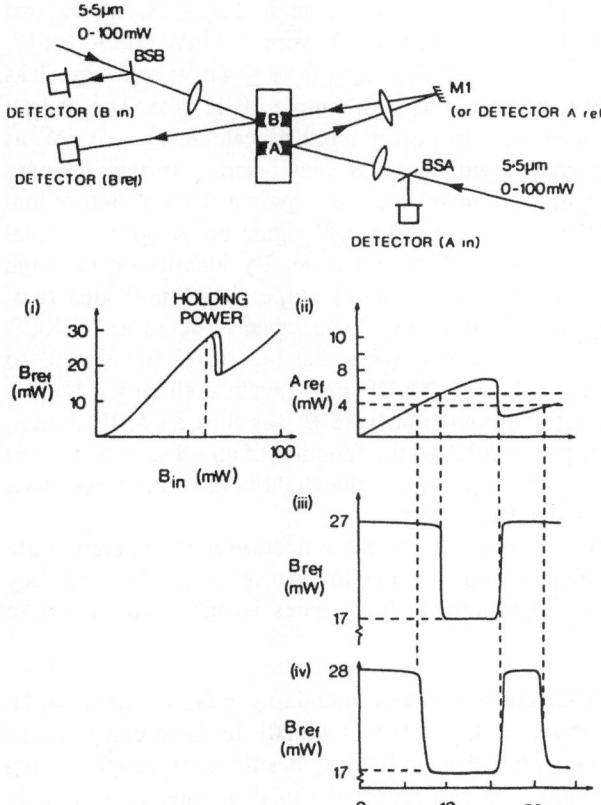

Fig. 4.18. *Top:* Schematic layout of the two-gate digital optical circuit formed using a single InSb nonlinear etalon. Curves (i) and (ii) show the response characteristics of the two gates B and A, respectively. Curves (iii) and (iv) show examples of the output of gate B as a function of the input to gate A, for two different holding powers on gate B. (Operating wavelength 5.5 μm)

(i) the two bistable elements were formed on a single InSb sample (500 μm apart); and

(ii) the *reflection* output from the first element was used to switch the second.

Figure 4.18 shows schematically the optical arrangement for this circuit, plus the output characteristics for each gate and the overall response of the combination. The reflected power from gate A was directed at gate B, which was held just below its switch point. When the input to gate A was steadily increased, the increase in reflected power was sufficient to switch on gate B. However, a further increase in input to gate A caused a switch to the low reflection state and, consequently, gate B switched off simultaneously. Finally, a further increase in the input power to gate A could switch gate B to the on state once more.

By adjusting the proximity of the bias level of gate B to its self-switch power, the output level from gate A necessary to cause a switch could be varied.

The performance of the coupled gates, as presented in Fig. 4.18, shows two examples in which the critical *outputs* from gate A were $\sim 4\,mW$ and $\sim 6\,mW$. Thus, taking the second example, by defining logic-0 as $\leq 4\,mW$ and logic-1 as $\geq 6\,mW$, gate B responded to these inputs as a simple NOR gate. The overall circuit also recognized an increment in power *input* to gate A of $\sim 6\,mW$ as a logic-1. Thus by holding gate A with a $6\,mW$ bias beam, a further increase in input due to a $6\,mW$ signal (to $12\,mW$) caused a switch from a high output ($28\,mW$) to a low output ($17\,mW$). A second $6\,mW$ signal beam (giving a total input of $18\,mW$) returned it to the high output state. By identifying the high ($28\,mW$) output as logic-1 and the low ($17\,mW$) output as logic-0, this two-gate circuit was performing an XNOR function; i.e., gate A acted as an XOR gate and gate B inverted it. Alternatively, when the bias level (no signal) to gate A was set to $12\,mW$, it acted as an XNOR gate (again assuming a logic-1 level of $6\,mW$), and the inverted output from gate B was thus an XOR output. Besides inverting the output, gate B fulfilled the function of an off-axis addressed discriminator: recognizing the logic-0 to logic-1 threshold in the output from gate A and restoring the signal to standard levels.

The unique characteristics obtained in on-axis reflection mode operation are particularly useful in providing a range of functional responses. The way they can be exploited in making programmable logic gates is discussed further in Sect. 4.4.2d.

(b) Coupled NLIF Device Circuits. Circuits containing pairs of ZnSe NLIF devices were described by *Smith* et al. in 1985 [4.43, 20]. In these experiments, separate NLIF elements, each with their own holding beams, were coupled using imaging lenses. Both the transmitted and reflected signal outputs were used in experiments that showed that when one gate was switched on, the difference in output was sufficient to change the state of the next gate. The layout, shown in Fig. 4.19a, took the transmitted output from gate 1 and directed it at gate 2 in addition to a holding beam. When gate 1 was switched onto resonance, gate 2 was also observed to switch. Alternatively, with the arrangement shown in Fig. 4.19b, the reflected output from gate 2 could be directed at gate 1. In this case, when gate 2 was switched onto resonance, the decrease in reflected power was sufficient to cause gate 1 to switch off. The latter circuit behaved in a manner similar to the coupled InSb gates described in the previous section. Thus, the transmitted and reflected outputs from the second gate behave as XOR and XNOR gates.

(c) A Parallel Flip-Flop Circuit. Given the equivalence between a reflection-output bistable etalon and the inverted response of a transistor (see Fig. 4.7), a range of optical circuits may be constructed that are directly equivalent to their electronic counterparts. For example, two coupled NOR gates, i.e. two reflection-coupled nonlinear etalons, can form an RS flip-flop [4.17].

Figure 4.20 shows a schematic of a circuit used to demonstrate the operation of four parallel RS flip-flops. The optical elements were NLIF/BEAT reflection

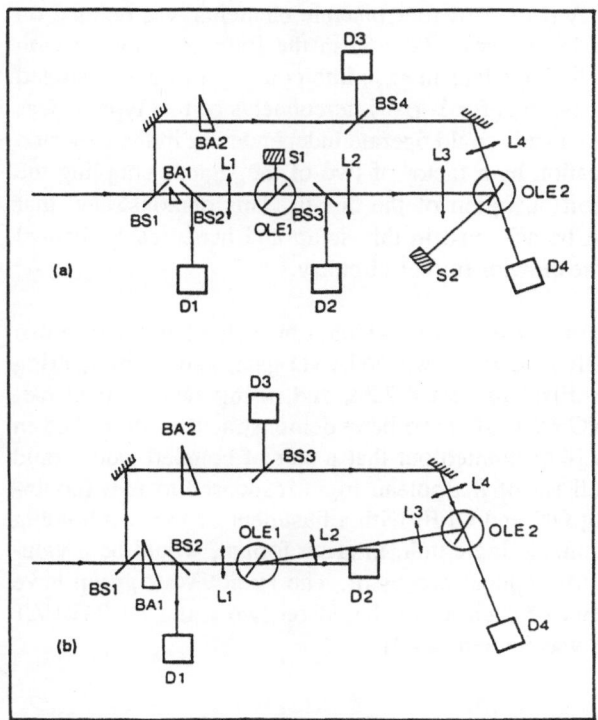

Fig. 4.19a,b. Schematics of optical arrangements used to demonstrate two-gate circuits based on ZnSe NLIFs. (a) Transmission output of optical logic element 1 (OLE 1) is used to switch OLE 2. (b) Reflection output of OLE 2 is used to switch OLE 1. BS: beam splitter; BA: variable beam attenuator; L: lens; D: detector. (Operating wavelength 514 nm)

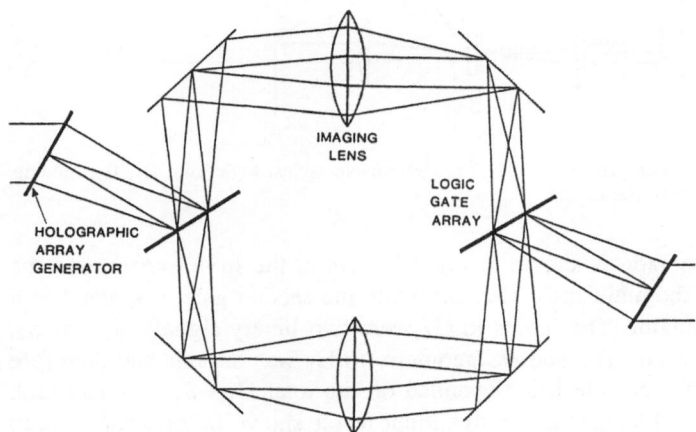

Fig. 4.20. Schematic showing two channels of a four-channel parallel optical flip-flop circuit. The logic gate arrays were ZnSe NLIF-BEAT devices operating in reflection. The circuit was powered by an argon laser (514 nm) via holographic array generator optics

devices working at 514 nm. A pattern of four bistable elements was defined on each plate by a holographic lenslet array (2 × 2) in the form of a square with sides 1 mm in length [4.34]. A single lens in each interconnect arm demonstrated the efficacy of such imaging systems for 1-to-1 interconnects of this type. It was shown that each of the four channels could operate independently in the expected manner. Furthermore, attenuation by a factor of two of the signals coupling the arrays together did not inhibit operation of the circuit. This demonstrated that a fan-out of at least 2 could be achieved in this setup and hence each channel could, in principle, communicate with further circuitry.

(d) Programmable Logic Units. The way in which a bistable etalon device can be changed from an OR/NOR gate to an AND/NAND gate, simply by altering the holding power, was described in Sect. 4.2.3d, and, using two coupled elements, the logic functions XOR/XNOR have been demonstrated as described in Sect. 4.4.2a [4.42]. *Wherrett* [4.6] pointed out that a pair of coupled gates could consequently provide the full set of 8 Boolean logic responses to two (noninverted) inputs (i.e., including ON and OFF) with adjustment of the two holding beams only. Such a programmable logic unit, in array format, would be a valuable component within a SIMD optical processor. The Heriot-Watt group have demonstrated a single channel of such a unit based on two ZnSe NLIF/BEAT devices operating at 830 nm wavelength [4.44].

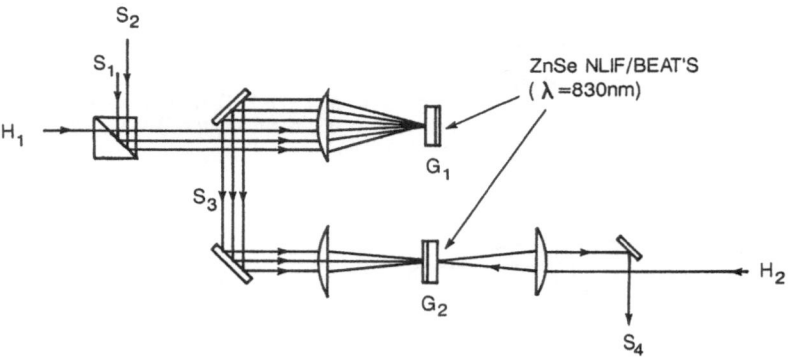

Fig. 4.21. Schematic of a two-gate (G_1 and G_2) programmable optical logic unit. H_1, H_2: holding beams; S_1, S_2: input signals; S_4: output

The circuit schematic is shown in Fig. 4.21. As in the InSb experiment, the first gate, G_1, was the main logic element while the second gate, G_2, acted as a logic level discriminator. The inputs to G_1 were two binary signals, S_1 and S_2, plus a holding beam H_1. The address geometry to G_1 was on-axis and therefore the reflection transfer charcteristic depended on the total $H_1 + S_1 + S_2$ and took the form indicated in Fig. 4.22a. The five input levels shown (a–e) correspond to logic-1 increments. The G_1 output, S_3, was directed onto the absorber layer of G_2 (off-axis address). The final output, S_4, was the reflected part of the holding beam H_2. The response of G_2 as S_3 was varied as indicated in Fig. 4.22b, for

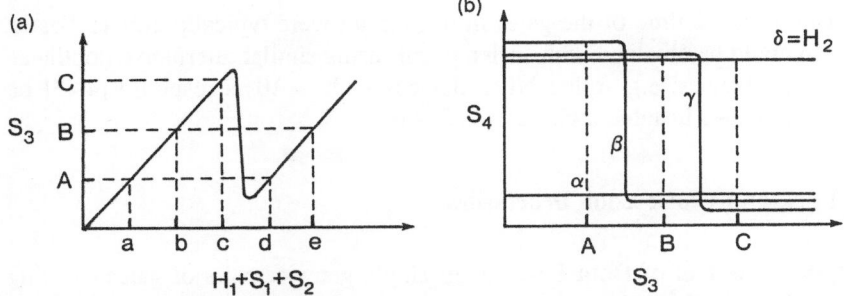

Fig. 4.22. (a) Idealized transfer characteristic for gate G_1 in Fig. 4.21. (b) Idealized response of gate G_2 in Fig. 4.21, to input S_3 for different holding powers H_2

four different values of H_2 (α–δ). The combination of these two characteristics produces all eight of the two-input logic functions: ON, OFF, OR, NOR, AND, NAND, XOR, XNOR.

In the experiments, a Styryl-9 cw dye laser supplied the input signals and H_1, and a cw GaAs diode laser provided H_2. In this initial study, only one signal beam, with 3 different levels [corresponding to the pairs of signal levels (0,0), (0,1)/(1,0), (1,1)] was incident on the first gate. This input was cross-polarized relative to H_1 to avoid interference problems.

The results are shown in Fig. 4.23. Using logic-0 $\leq$ 1 mW and logic-1 $\sim$ 3 mW, all eight logic functions were demonstrated with well-defined out-

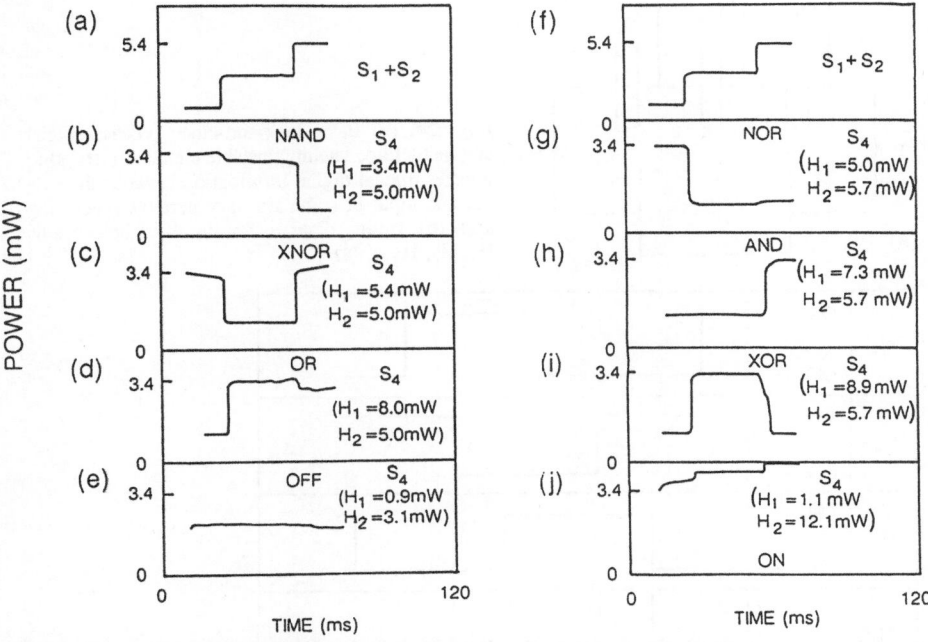

Fig. 4.23a-j. Results obtained from the programmable logic unit of Fig. 4.21. The different logical responses to the three possible input values for $S_1 + S_2$ were selected by setting H_1 and H_2 as indicated

puts. The response time of the gates in this circuit were typically $100\,\mu s$. Faster operation could be obtained with such a circuit using similar alternative nonlinear Fabry-Perot devices: e.g., other NLIF devices with $\sim 10\,\mu s$ response [4.45] or nanosecond GaAs bistable switches [4.23, 46].

4.4.3 Lock-and-Clock Loop Processors

(a) Optical Control of Data Flow. After single gates or pairs of gates carrying out logical operations on optical inputs had been demonstrated, the next step was to consider the way in which optical data might be transported in a controlled fashion through a more complex digital optical computer. This immediately raises a need for local temporary stores. The use of a single optically bistable gate as a volatile memory was described in Sect. 4.2.3e. By placing three such latching gates in sequence, it is possible to store separate signals (or in 2d, two images) in the outer devices, while using the central device as a shutter or buffer to avoid contamination between the information planes (Fig. 24a), [4.6, 7].

This is accomplished by setting the holding beams for the outer gates within the bistable regime (enabled) while disabling the central gate by holding it at a lower power. The extension to a chain of store-buffer pairs is straightforward. Such a multistage memory could be operated as a shift-register by sequential phasing of the holding beam levels such that information is passed on to a gate

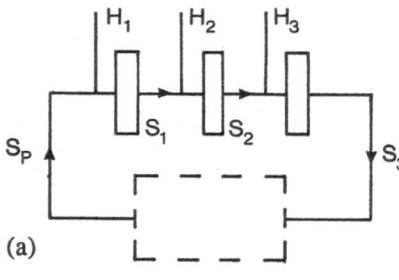

(a)

Fig. 4.24. (a) Basic lock-and-clock processing circuit using three latching bistable elements (arrays) to transfer optical digital information between the output and input of some arbitrary iterative processing unit. (b) Timing diagram for the three hold beams H_1, H_2, H_3 of (a)

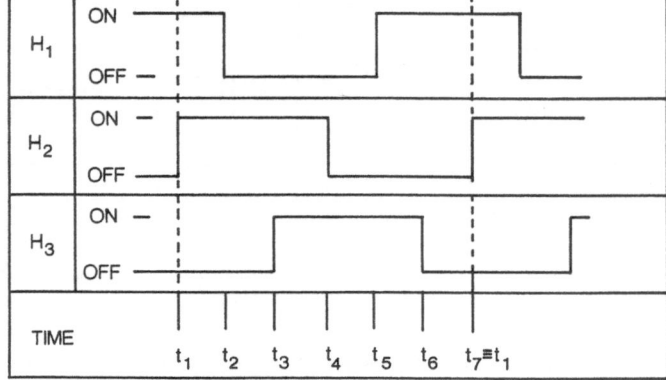

(b)

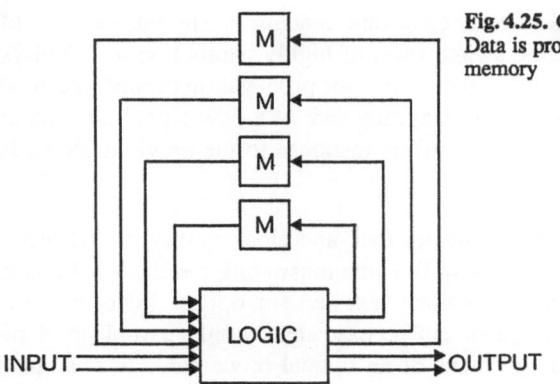

Fig. 4.25. Classical finite-state-machine concept. Data is processed iteratively via a parallel-access memory

only when its bias is switched to enable and the next gate is disabled. In its 2d form, this is a parallel shift register with an optical wiring harness.

By recycling the output of such a register through a processing unit and back to the input, a classical finite-state-machine architecture is achieved (Fig. 4.25). This is the basis of many of the computing architectures that aim to overcome the Von Neumann bottleneck of conventional sequential machines. With such architectures, data is circulated in parallel between the processor and a latching store, rather than data being acquired sequentially in individual units from a random access memory using an address technique. It is precisely this architectural style that favors an optical approach. In its simplest form, the memory in Fig. 4.25 is a sequence of just three 2d parallel bistable arrays (Fig. 4.24a). At a certain stage of the control cycle, the output array stores a preprocessed image, the input array stores the processed version of this image, and the central array is switched off. The time sequence of enabling the three logic arrays in order to control the flow of information around the circuit is shown in Fig. 4.24b. Because this time-sequencing leads to the locking of information on one array, followed by clocking onto the next, it has become known as a *lock-and-clock* circuit.

Further properties of the lock-and-clock circuit include the following:

(i) Synchronization across each 2d array is provided by the control beams, so that the logic unit receives and processes planes of information in a lock-step fashion.

(ii) With proper use of the bistable switches, logic level restoration is achieved, ensuring that over many cycles there is no accumulation of small errors. Thus, provided that the first bistable array distinguishes correctly between low and high power output signals from the processor stage correct and standardized inputs at the next processing cycle are then provided by the response of the third array. This is a very important feature of the architecture.

Various alternative bistable loop circuits have been described [4.6, 7], such as that presented in concept in Sect. 4.1.1 (Fig. 4.2). The synchronization unit may be thought of as a controllable *optical time delay*. It operates rather like the

125

acoustic delay lines introduced in early electronic machines. The introduction of the delays is essential because the switch rates of highly parallel optical devices are necessarily long compared to the speed-of-light propagation around practical circuits. By introducing synchronized delay, the risk of a new signal arriving at the input to a gate before it has completed its response to the previous signal is avoided.

(b) The Basic Loop Circuit. The simplest lock-and-clock system is the basic loop circuit formed by three gates. As well as demonstrating the lock-and-clock concept, such a circuit acts as an excellent test-bed for optical logic devices. Demonstrating that optical logic gates can be cascaded is only partial proof of their ability to act as the building blocks of an optical processor. For example, a sequence of increasingly sensitive gates may successfully pass a signal along the chain, but the final element would be incapable of switching the first. This situation we describe as "down-hill" switching which is clearly undesirable. It is necessary to demonstrate *restoring logic*, in which the levels are restored to a standard level, either at each gate or at the output of simple modules consisting of a few coupled gates. Given that there are inevitable losses due to signal fan-out and attenuation during transmission, this corresponds to the requirement for gain discussed in Sect. 4.2.3.

The first lock-and-clock circuit that was operated [4.47–49] is shown schematically in Fig. 4.26a. It was constructed using three bistable ZnSe NLIF devices, each acting as just a single transmission-output switch with off-axis address. The circuit demonstrated the input and repetitive circulation, without error, of a single bit of data.

As described in the previous section, the lock-and-clock technique proceeds by stepping the holding beams between near switch point and zero, sequentially enabling and disabling each gate. Thus, in this experimental circuit, a particular bistable element:

(i) was enabled, so as to respond to a transient input (e.g. from the previous gate);
(ii) by latching, continued to store this datum when the input was removed;
(iii) passed on this datum to the next gate when that was brought from the disabled to enabled state; and
(iv) was reset as it was disabled.

This corresponded to a three-phase clock sequence for the holding beams, in which no more than two gates were enabled at any one time. Gate A, with an input from the previous gate plus an external signal, acted as a two-input OR-gate and gates B and C as one-input OR-gates. The three elements were powered by an argon-ion laser operating at 514 nm, via beam-splitters, static attenuators, and computer-controlled acousto-optic modulators (AOMs). The latter were used to impose the three-phase holding-beam clock sequence and acted as an interface with electronic control. The external signals were determined by a fourth AOM. Figure 4.26b shows the waveforms recorded at the outputs of each gate (via beam-

(a)

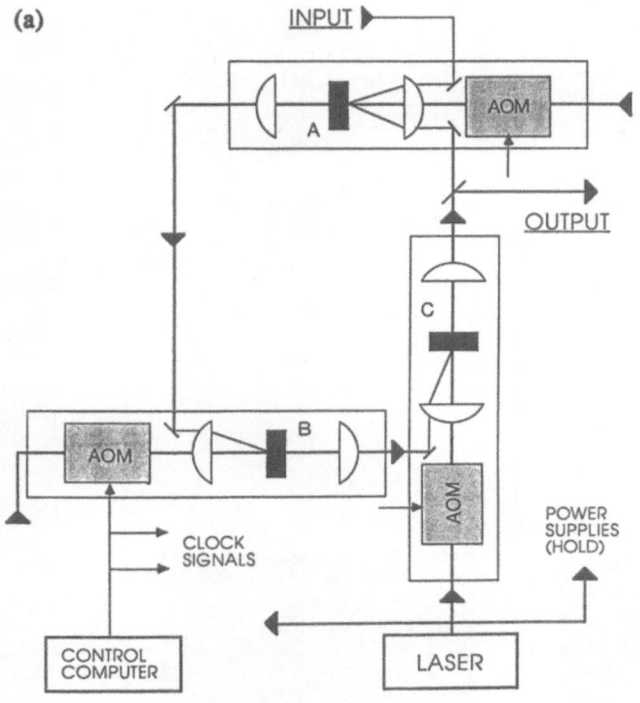

(b)

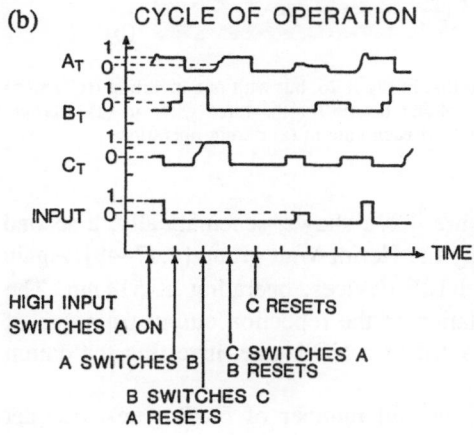

Fig. 4.26. (a) Schematic of the first experimentally demonstrated lock-and-clock circuit. One bit of data could be input to gate A and, via gates B and C, passed repetitively around the loop. The gates were ZnSe NLIF devices powered by a 514 nm argon laser via control acousto-optic modulators (AOMs). (b) Output sequences from the three gates in (a). In this example a logic-1 was circulated, followed (after a reset) by a logic-0 and another logic-1

splitter/detector combinations) as the circuit operated. A logic-1 was propagated around the loop in synchronism with the clock signal, followed on the next loop cycle by a signal that was (intentionally) too small to switch (i.e. logic-0), which left the three gates in their logic-0 states when enabled. (All gates were reset to zero before each cycle).

127

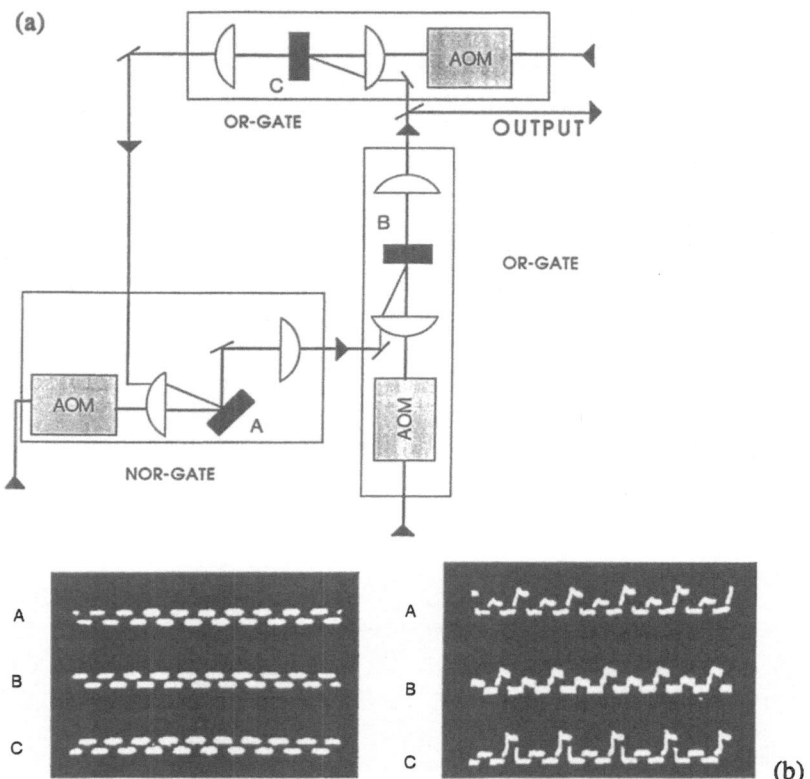

Fig. 4.27. (a) A similar lock-and-clock circuit to that in Fig. 4.26, but with one inverting (reflection) gate (A). (b) *Top*: The three-phase (clocked) holding beams for the three gates in (a). *Bottom*: Waveforms measured at the (transmission) outputs of each gate in (a) during operation

(c) Loop Circuit with Inversion. Figure 4.27a shows, schematically, a second loop circuit that was also constructed by the Heriot-Watt group [4.47–49]. Again it was based on three bistable ZnSe NLIF devices, operating at 514 nm. The difference in this case was the exploitation of the reflection output from one of the gates. This element consequently acted as a NOR-gate, inverting the datum on each cycle.

A circuit of this type, containing an odd number of NOR gates, can act as a free-running digital oscillator when all elements are held steady in their enabled states [4.17]. This mode of operation was demonstrated experimentally but exhibited a very poorly defined oscillation frequency due to the sensitivity of gate switch times to small fluctuations in input power and additional problems due to switch power drift (Sect. 4.3.3). Better performance was achieved using lock-and-clock control, as demonstrated with the previous loop circuit. Because this circuit automatically inverted a logic-0 to logic-1, no external input was required to test for correct operation.

Figure 4.27b shows both the three-phase clock sequence applied to the three gates and the detected (transmission) outputs. It can be seen that, because the signal is inverted at gate A, the response of gate B is in the opposite sense. Thus the single bit of information circulating in the loop changed from logic-0 to logic-1 and back again on each consecutive cycle.

These simple circuit experiments demonstrated

(i) that restoring optical digital logic systems can be constructed from bistable devices and
(ii) that the negation operation, essential in any digital computation system, can also be realized in practice using the reflection output.

(d) Parallelism – The Optical Classical Finite-State Machine. The above lock-and-clock circuits, operating with just a single switch on each bistable logic plane and only one bit of information in the loop, are valuable, but at the same time very limited, demonstrations. They involve more control electronics than they contain optical logic! As has already been indicated, such arrangements *will* be of computational interest when they support many thousands of channels circulating in parallel (with suitable interconnects). As a small step towards such a parallel configuration, a lock-and-clock circuit has been constructed with three parallel channels.

Figure 4.28 shows the arrangement schematically. Once again the system operated with 514 nm light but in this case each logic plane operated in the reflection mode and consisted of a ZnSe NLIF/BEAT device with an evaporated-

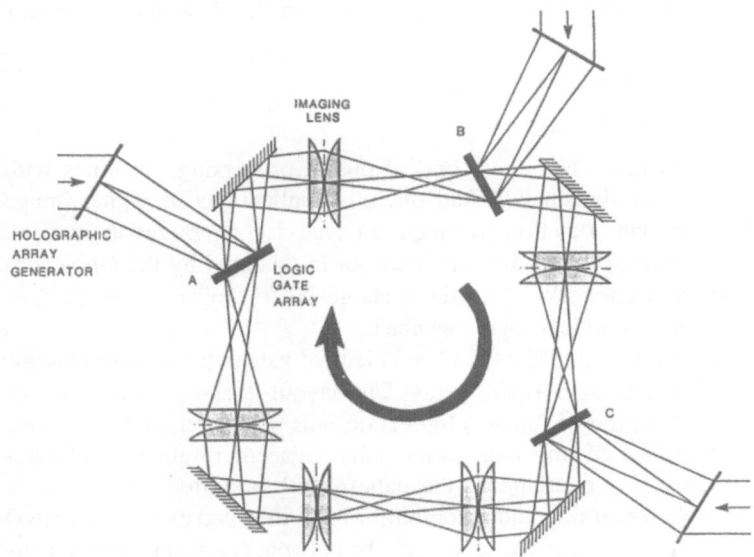

Fig. 4.28. Schematic of a parallel lock-and-clock circuit. In this case, an array of three bistable switches was operated on each logic plane: A, B and C. These were ZnSe NLIF-BEAT devices operating at 514 nm

silicon absorbing layer. The threefold parallelism was achieved by illuminating an array of points on each (uniform) bistable plate using separate holding beams. Each gate was separated from its neighbor by 2 mm and acted as an independent switch. In a similar way as in the parallel flip-flop circuit described in Sect. 4.4.2c, the holding beams were produced by an array of dichromated gelatin (DCG) holographic lenslets. In this experiment, they were arranged in a triangular pattern so as to generate corresponding sets of focal spots when illuminated by a uniform-irradiance parallel beam. A condenser lens imaged the reflected (output) signals onto the input plane of the next array: the silicon absorber layer. This same interconnect would have been equally compatible with a much greater level of parallelism: in principle, the holographic array generator would just need to be replaced with another producing more holding beams.

The system was again clocked in the three-phase sequence by AOM control of each of the three power beams *before* they were fanned-out by the lenslet array. Switching of each array of optical logic gates by the outputs from the preceding array was demonstrated. In this way, it was shown that parallel all-optical digital logic systems with spatially invariant interconnects can be constructed using nonlinear (bistable) Fabry-Perot devices.

The three bistable arrays correspond to the buffered memory unit discussed in Sect. 4.4.3a and shown in Fig. 4.24 [4.6]. With such systems, a complete image, corresponding to a 2d binary data field, could be processed iteratively, as discussed further in Sect. 4.4.4d. Alternatively, by comparing Fig. 4.28 with Fig. 4.25, it can be seen that this circuit had many of the features of a classical finite-state machine. Two of the gate arrays could be regarded as memory, into and from which data can be clocked, and the third array as a (clocked) logic unit. In this case, a simple logical inversion was applied to the data before directly recycling.

4.4.4 Image Processing

(a) **Pattern Recognition.** Nonlinear optical digital processing machines with very high levels of parallelism will find obvious applications in digital image processing. An important function in image analysis is pattern recognition. A series of initial experiments have been carried out in this area by the University of Arizona group and co-workers. Again, bistable NLIF devices were used as convenient, available all-optical logic switches.

The first demonstration [4.50] used a 2×3 element gate pattern in conjunction with a graded-index-lens input-optics array. The layout is shown schematically in Fig. 4.29. A 2×2 pattern of (binary) light spots was presented, and the system recognized the presence of any two horizontally adjacent bright spots (logic-1). This was achieved by overlapping the pattern with a shifted copy of itself (implemented by beam-splitting and recombining) and then carrying out an AND operation. The output corresponded to a single bright spot (logic-1) wherever the pattern was recognized. The system was clocked so as to reduce total thermal dissipation requirements and to ensure re-set of the bistable gates. The optics

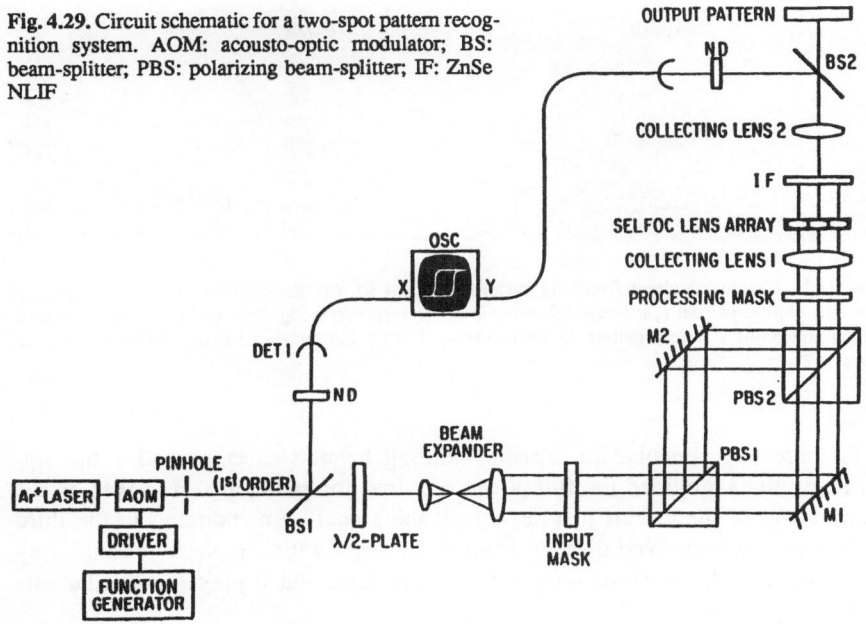

Fig. 4.29. Circuit schematic for a two-spot pattern recognition system. AOM: acousto-optic modulator; BS: beam-splitter; PBS: polarizing beam-splitter; IF: ZnSe NLIF

providing the shift interconnect could have equally well handled a much larger array.

The next step in complexity is to recognize a 3-spot pattern, and this was demonstrated by the Arizona group using cascaded bistable NLIF arrays with a 2×9 input pattern [4.51]. A schematic of the optical arrangement is shown in Fig. 4.30. The system operated by first recognizing the presence of two of the spots in the required pattern, in the manner just described, and then carrying out a further AND operation between the output from this stage and the signal corresponding to the position for the third spot in the pattern. Examples showing operation of this circuit are presented in Fig. 4.31.

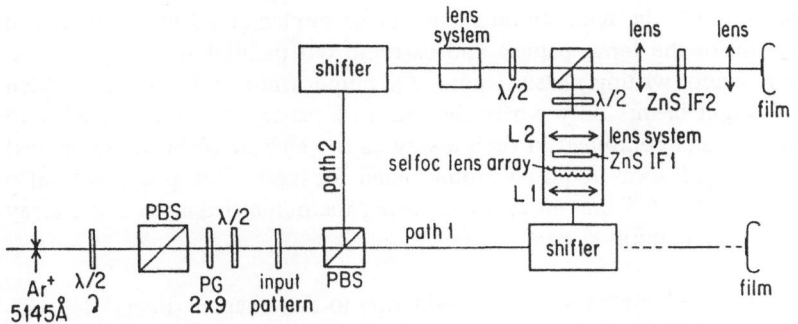

Fig. 4.30. Circuit schematic for a three-spot pattern recognition system, based on two 2×9 ZnS NLIF arrays (IF1 and IF2). PBS: polarizing beam-splitter; PG: phase grating

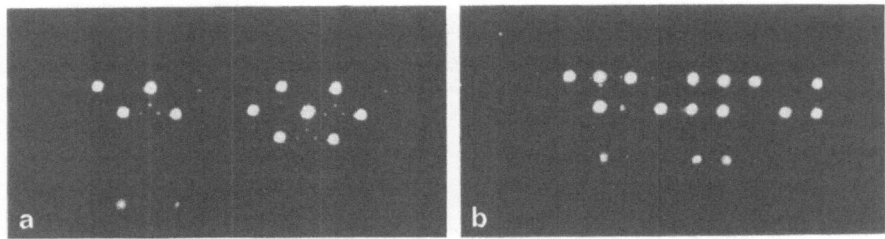

Fig. 4.31a,b. Results obtained from the system in Fig. 4.30. (a) Recognition of V. Input pattern *upper left*; shifted pattern *upper right*. Result, *lower left*, shows recognition of the single V pattern. (b) Recognition of Γ. Input pattern *upper two rows*. Result, *bottom row*, shows correct recognition of three Γ's

In these experiments, no separate holding beam was employed – the signals themselves provided the full power required for switching. However, in the second stage of the circuit just described, the signal corresponding to the third required spot was derived directly from the laser, via binary phase grating array generators plus the mask defining the input pattern, and it played a similar role to a holding beam.

These experiments demonstrated that all-optical bistable devices can indeed be used in pattern recognition systems in conjunction with space invariant interconnects.

(b) Symbolic Substitution. Symbolic substitution was proposed by *Huang* [4.52] as a parallel algorithm approach to the construction of high parallelism optical computers. The technique proceeds by recognizing a pattern and then replacing it with some (pre-specified) alternative pattern. This is done in parallel over a large data field within which there may be many such instances of recognition and substitution. By embedding such a processor within an iterative loop, a wide variety of processing operations could be carried out, including digital numerical calculations.

The simple demonstrator circuits described in the preceding section (Sect. 4.4.4a) showed how the recognition stage can be implemented by combining a shifted version of the input pattern and carrying out parallel AND operations. The second symbol writing phase requires the output from such a stage – which consists of bright beams only where the required pattern was recognized – to be split, shifted and combined in such a way as to generate (from these beams) the new pattern of spots required around each location. This phase was also demonstrated by the Arizona group [4.50] using the output from the 2×3 array 2-spot pattern recognition system.

(c) A Digital Edge Extractor. The introduction to this chapter (Sect. 4.1.1) discussed different classes of optical processing, each identified by varying degrees of fan-out. The lock-and-clock circuits and the simple pattern recognition systems

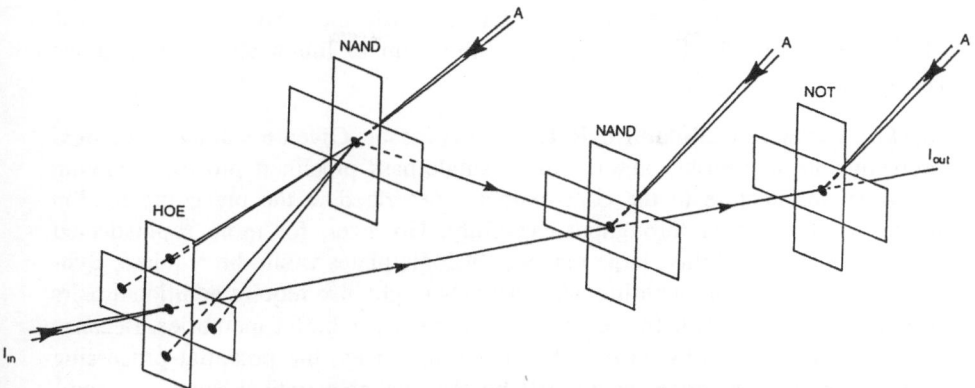

Fig. 4.32. Schematic of an elementary cellular logic image processor, designed to perform digital edge extraction on a binary image. This pipelined processor was constructed using a DCG-holographic 1-to-4 fan-out element and ZnSe NLIF devices. (514 nm operation)

described in Sects. 4.4.3, 4.4.4a and 4.4.4b fall into the first category, relying on minimal fan-out levels, i.e. 1-to-1 or 1-to-2. The second class that was identified covered cellular architectures for digital image processing which typically require fan-out levels of 1-to-4 up to about 1-to-8 (e.g. for nearest-neighbor interconnects). Whilst these can be built up, in principle, by combining 1-to-2 fan-out stages, it is of interest to test the higher fan-out, potentially more efficient, architecture. To investigate the feasibility of implementing such image processing architectures optically, the Heriot-Watt group designed and tested experimentally a digital edge extractor, again based on nonlinear Fabry-Perot devices.

The system is shown schematically in Fig. 4.32. To recognize edges on a binary image, it is necessary to identify those bright pixels that have one or more dark neighbors. This can be done by fanning out the signal from each pixel to its four nearest neighbors, implementing a four-input NAND operation on the resulting image, and finally carrying out an AND operation on this result with the original image. Because the output of the NAND stage is logic-1 only when one or more of the four nearest neighbors (that fan-in together in this plane) is logic-0, it follows that an output from the following AND stage is only obtained when a logic-1 pixel is next to a logic-0 pixel, i.e. at the edge of the image.

The experimental test of this pipeline image processing system was again implemented only in its simplest form. Essentially, a single logic channel was tested but with a pixel plus four nearest neighbors as inputs, as indicated in Fig. 4.32. The logic gates were ZnSe NLIF/BEAT devices operating at 514 nm in the reflection mode (off-axis address). Because of the inverted nature of the reflection gates, an additional stage was required in order to obtain the AND operation.

A 1-to-4 fan-out hologram was made for the nearest-neighbor interconnect [4.53] required in this system. It was a spatially invariant element designed to operate on the Fourier transform plane of a (square) optical array so as to yield nearest-neighbor interconnects when re-transformed. Such an interconnect element could be used on any size of array, up to limits set by its angular acceptance.

(d) The Iterative Cellular Logic Image Processor. Given a suitably presented problem and appropriate algorithms, a single-pass pipelined processor having a similar architecture to the edge-extractor described in the preceding section provides an optimum throughput capability. However, for more sophisticated processing and flexibility, large numbers of logic plates would be required. Synchronization between incoming data channels might also require additional pairs of latching gate arrays before each processing unit, as buffer memories. Because each gate array requires its own holding beam array, the potential processing power would, in practice, be limited by the available optical power. A compromise is the iterative optical cellular logic image processor (optical CLIP) architecture, which trades off some processing rate against a reduction in the numbers of logic planes, and hence power consumption, while retaining flexibility in processing capability [4.54]. This architecture is a combination of the, already demonstrated, iterative lock-and-clock approach (Sect. 4.4.3) with a programmable logic unit (Sect. 4.4.2d) used in a manner analogous to the logic planes in the edge-extractor system (Sect. 4.4.4c).

The optical CLIP architecture is based on a 2d array of processing cells each receiving two inputs and producing two outputs. Temporarily, one output is the signal of interest, the other is fed as an input on the next clock period to each of a set of other (neighboring) processing cells. Each cell receives one basic input plus a second that was generated during the previous cycle from the cells for which it is a neighbor (Fig. 33a). The term "neighbor" is used here in a general sense; it could refer to nearest neighbors in the array, to nearest plus next nearest, to a group of cells separated by some distance from the one in question or to some dispersed set of cells. The signal derived from the neighbor inputs can be a simple thresholded signal. Figure 4.33b shows a schematic of a possible configuration for the whole system. F_N and F_A would be programmable logic units. Thus F_N represents a 2d array of processing elements that would produce those signals that are to be passed to neighboring elements, and F_A would produce the set of output signals (the output image at a given time period). F_N and F_A receive the same two-input 2d images. To achieve the neighbor interconnects simultaneously over the entire array, a space invariant interconnect would be used, followed by a thresholding logic-gate array.

We see a computational system of this sort as the next step in developing the optical digital image processing concepts described earlier in this chapter. Using nearest-neighbor interconnects only, a number of image algebraic processes could be achieved by suitable choice of the combinatorial logic functions employed at F_N and F_A: e.g. expansion, compression, skeletonization, noise removal, edge-extraction, etc. More general physical problems, such as the Ising model (discussed in Sect. 4.1.1) could also be tackled in 2d with binary data fields input from an electrically addressed spatial light modulator. Figure 4.34

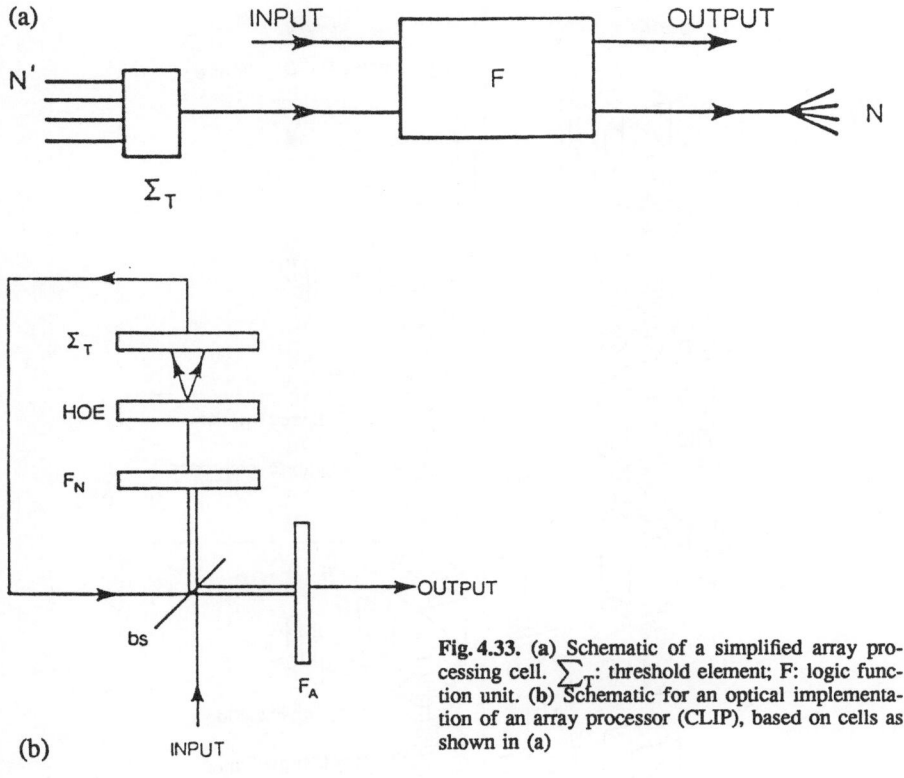

Fig. 4.33. (a) Schematic of a simplified array processing cell. $\sum_T$: threshold element; F: logic function unit. (b) Schematic for an optical implementation of an array processor (CLIP), based on cells as shown in (a)

gives an indication of the optical component layout that might be required for óne possible implementation of the optical CLIP concept.

(e) Associative Memories. As noted in Sect. 4.1.1, the class of architecture with the highest levels of optical fan-out are those based on Fourier processing concepts. Because of the large fan-in of signals in the Fourier plane, this type of optical computation is generally limited to analog processing. The experiment described in this section is one in which such an analog system is combined with a nonlinear digital optical switch (an NLIF device) which acts as a binary decision maker. The experiment was based around a holographic associative memory concept and was carried out as a collaboration between *Khitrova* et al. at Arizona University and *Psaltis* at Caltech [4.55, 51].

Associative memories permit the recall of complete sets of data from an input of only part of the data (within limits set by the degree to which the partial input is incomplete and the inevitable presence of ambiguities). In an optical context, each complete data set could be an image. If the partial input, e.g. half an image or a noisy version of it, results in the recall of the complete original, this is known as an auto-associative system. Alternatively, the partial input could stimulate the output of an entirely different image, which is a hetero-associative process.

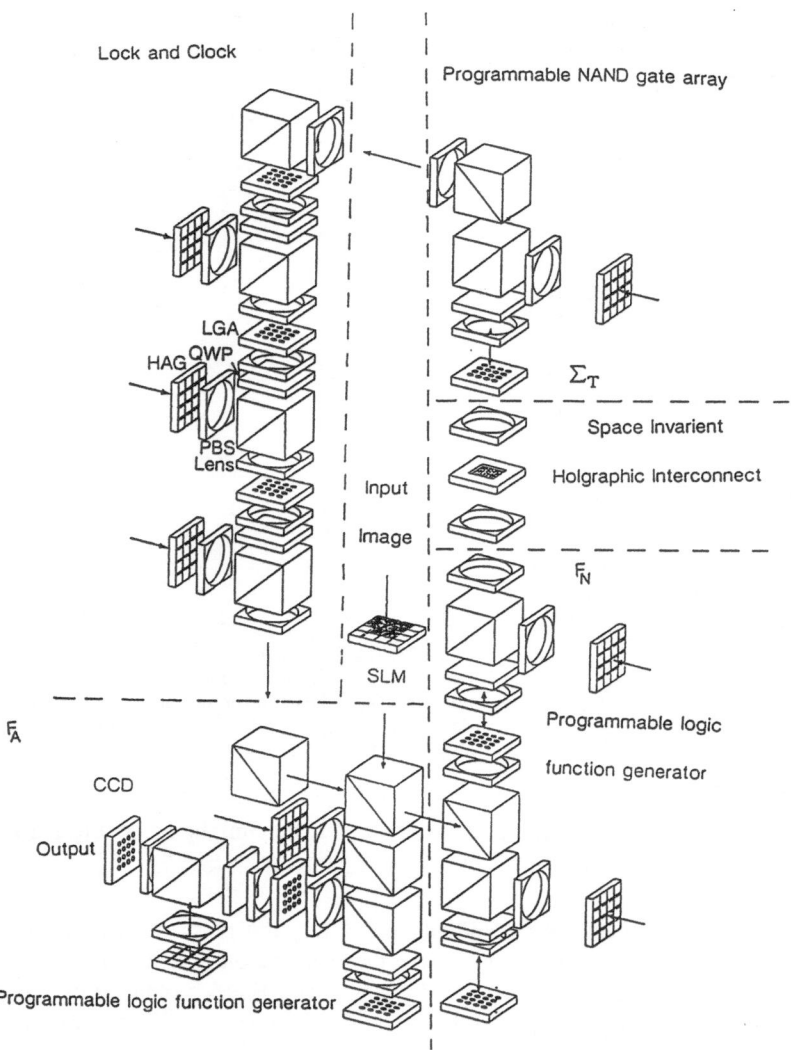

Fig. 4.34. A possible optical layout for implementation of a CLIP. SLM: spatial light modulator; HAG: holographic array generator; QWP: quarter-wave plate; PBS: polarizing beam splitter; LGA: logic gate array

Associative optical memories can be implemented using holographic techniques. The simplest example of a hetero-associative system is a hologram formed by the light from two objects. Each acts as a reference beam for the other, such that on replaying with the light from just one of them, an image of the other is formed. More usefully, a series of holograms recorded within a single volume holographic medium can be addressed by the partial input of any one of them. If each hologram is recorded with angularly separated reference beams, then an output beam is generated in a direction corresponding to the reference beam that is associated with the best matched image [4.56]. Because there is always some degree of correlation with the other stored images, a nonlinear element can be included to recognize which of the outputs is dominant. Having selected the correct output, it can be used either to read out the complete stored image (auto-association) or some entirely different image (hetero-association).

In the experiment described here, overlaid Fourier plane holograms of two different fingerprint patterns were recorded in dichromated gelatin (DCG). (By working in the Fourier plane, the arrangement corresponded to a van der Lugt correlator, which has the advantage of being insensitive to lateral translation of the input image.) A schematic of the optical layout is shown in Fig. 4.35. The nonlinear optical switch was a ZnS NLIF device, biased near switch point by an independent holding beam counterpropagating relative to the signal coming from the hologram. When an input image produced a sufficiently strong correlation output in the direction of the NLIF element, the holding beam was switched through onto the hologram at just the correct angle to read out the full image (auto-association). Alternatively, this switched beam could be equally well used to read out a different image from a second hologram (hetero-association).

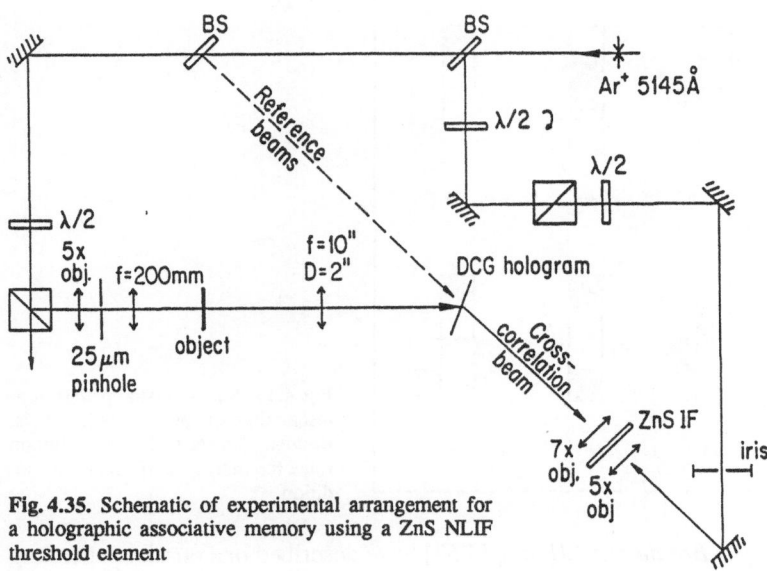

Fig. 4.35. Schematic of experimental arrangement for a holographic associative memory using a ZnS NLIF threshold element

4.4.5 Numerical Processing

In addition to image processing systems based on parallel digital optical circuits, which would appear to be a natural application, it is worth considering the numerical processing potential of this technology. The fundamental basis of any digital arithmetic unit is a binary full-adder. If parallel full addition can be demonstrated, it is relatively straightforward, conceptually, to extend the architecture to achieve bit-slice (or word parallel) multiplication and, by incorporation of optical data reordering schemes such as the perfect shuffle, to implement such algorithms as the fast Fourier transform.

The following sections describe two possible approaches to parallel binary addition using 2d optical logic arrays.

(a) Addition by Symbolic Substitution. As discussed in Sect. 4.4.4b, symbolic substitution involves the processing of encoded patterns by first recognizing a particular arrangement of bright pixels and then substituting each occurrence of this pattern within a 2d data field by a second pattern. Although it is fundamentally an image processing concept, by using properly encoded data and suitable algorithms, numerical processing can be carried out in this way.

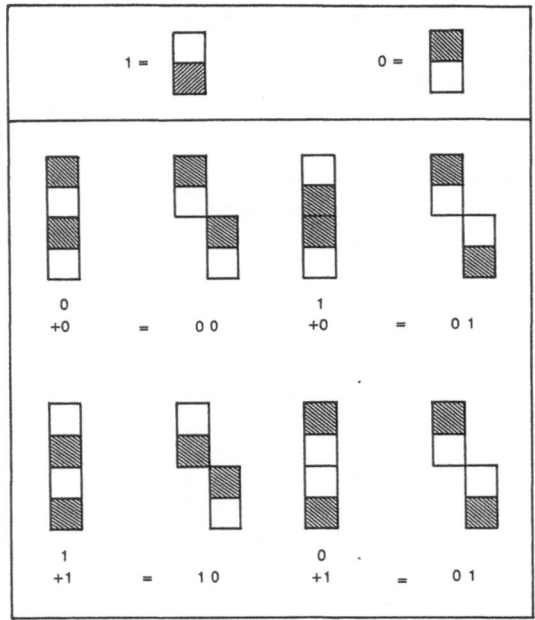

Fig. 4.36. *Top*: A spatial pattern representation of binary logic levels. *Bottom*: The set of four substitution rules required to implement full addition

For example, *Brenner* and *Huang* [4.57] have described one possible encoding and set of substitution rules, as shown in Fig. 4.36, which would permit addition of two binary numbers. Binary 0 and 1 are defined spatially by the position of a bright pixel within a pair of complementary pixels representing each binary digit. (This type of spatial encoding also has the advantage of always dissipating the same total power over the array and hence avoiding the varying thermal loads associated with single pixel binary representations.) The output pattern corresponds to the sum bit in the original position plus the carry bit, displaced by one digit.

The Arizona group have investigated the experimental implementation of a one-bit adder based on this approach, using ZnS NLIF logic devices operating at 514 nm [4.58]. A 1 × 4 array of beams, focused onto the NLIF by a fly's-eye lens array, provided the four pixels (1 mm separation) representing the two bits of binary data. The basic operations required to implement the sum rules correctly were demonstrated individually, but the complete one-bit adder was not constructed.

(b) A Nonlinear Etalon Full-Adder. It has already been noted that the on-axis addressed, reflection response of a nonlinear Fabry-Perot etalon device can provide a range of logic responses [4.6] (Sect. 4.4.2d). Used in conjunction with a level discriminator on the output, responses such as those in Fig. 4.18 can be obtained. *Wherrett* [4.59] pointed out that these correspond to the sequence of levels required for the sum output of a 3-input binary adder, while the transmission characteristic provides the response needed for the carry output. Figure 4.37 shows typical reflection and transmission responses for such a nonlinear etalon device. It can be seen that if the logic-0 and logic-1 outputs are discriminated as indicated, and a logic-1 corresponds to one unit, then for four total input levels as shown ($H, H + 1, H + 2, H + 3$), the reflection output response (0, 1, 0, 1) is indeed the correct sum response and the transmission output (0, 0, 1, 1) the correct carry response. (The importance of handling three binary inputs when adding just two binary numbers lies in the need to include the carry from the next significant place.)

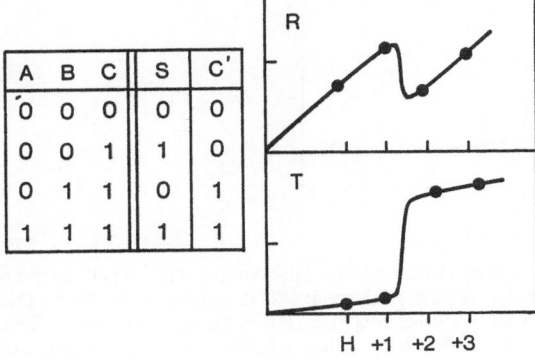

A	B	C	S	C'
0	0	0	0	0
0	0	1	1	0
0	1	1	0	1
1	1	1	1	1

H +1 +2 +3

Fig. 4.37. *Left*: Truth table required for a three-input (two bits, A and B, plus a carry, C) binary full-adder. *Right*: Transfer characteristics for a nonlinear Fabry-Perot etalon showing the response required for optical implementation of the full-adder. Reflection output, R, corresponds to the SUM, S. Transmission output, T, corresponds to CARRY, C'

Figure 4.38a shows, schematically, an experimental realization of this concept demonstrated by the Heriot-Watt group using ZnSe NLIF devices operating at 514 nm [4.60]. The system consisted of an on-axis addressed NLIF, the adder element, plus two NLIF/BEAT devices, the discriminator and level restoration elements, fed by the reflection and transmission outputs of the first element. Figure 4.38b shows the resulting outputs as a function of the total input to the adder plate. Because the discriminator elements operated in reflection mode, these outputs were, in practice, the inverse of the required response, but this could be easily corrected by an additional NOT stage.

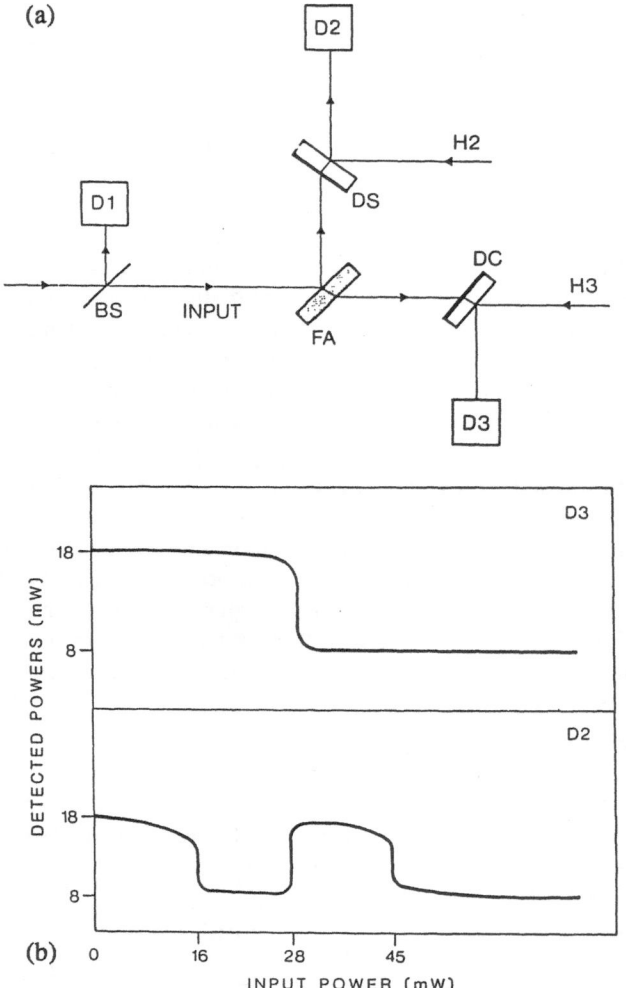

Fig. 4.38. (a) Schematic of optical layout used to demonstrate full-addition, with output discrimination to produce standard signal levels. FA: full-adder element; DS: discriminator for SUM; DC: discriminator for CARRY; *H*: holding beams; BS: beam splitter; D1–3: detectors. FA was a ZnSe NLIF and DS and DC were similarly NLIF/BEATs, all operating at 514 nm. (b) Output of two discriminator gates in (a) with increasing input to adder element, showing required CARRY (*top*) and SUM (*bottom*) response

Any such binary adder element must be capable of iterative operation; that is, the output of the carry channel has to act as an input at the next significant place on the next cycle, and the sum output must be able to drive any subsequent adder circuit. To demonstrate how this may be achieved experimentally, the circuit shown schematically in Fig. 4.39a was constructed. One of the outputs (the sum in this case) was looped around to act as one of the three inputs to the adder element. This could have been done by the standard lock-and-clock approach

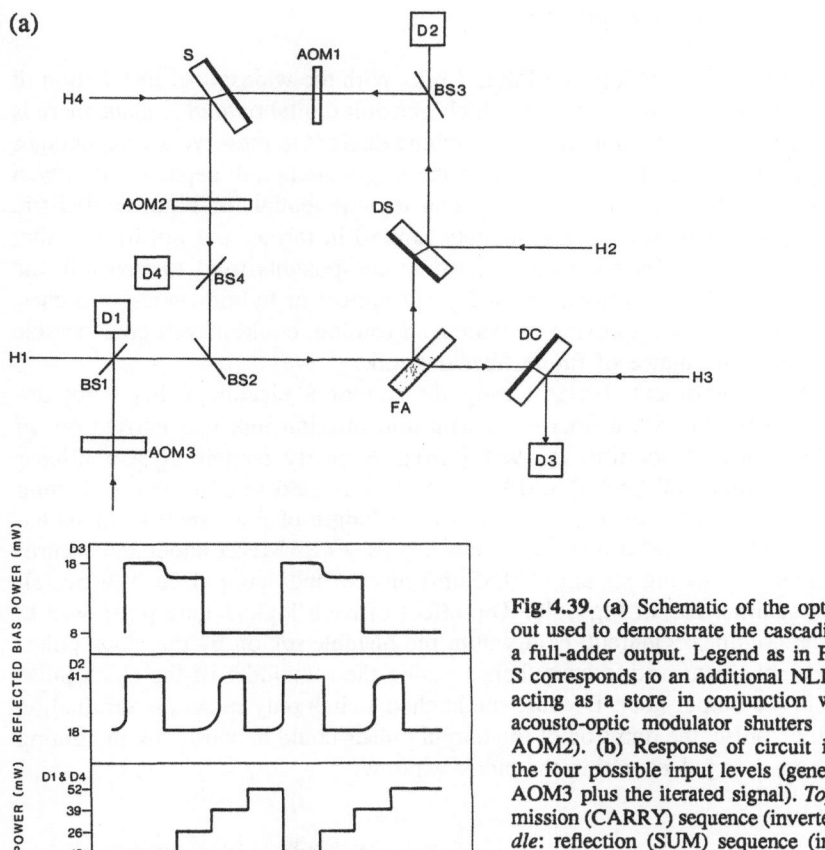

(a)

(b)

Fig. 4.39. (a) Schematic of the optical layout used to demonstrate the cascadibility of a full-adder output. Legend as in Fig. 4.38. S corresponds to an additional NLIF/BEAT acting as a store in conjunction with two acousto-optic modulator shutters (AOM1, AOM2). (b) Response of circuit in (a) to the four possible input levels (generated by AOM3 plus the iterated signal). *Top*: transmission (CARRY) sequence (inverted); *middle*: reflection (SUM) sequence (inverted); *bottom*: input sequence (total)

using three further bistable elements, as described in Sect. 4.4.3. For experimental simplicity, a single NLIF/BEAT element plus two acousto-optic modulators (AOMs) acting as clocking shutters were sufficient for this demonstration. (When reflection bistable elements are utilized, either approach inverts the signal in the manner required.) Figure 4.39b shows the correct variation in output of the two adder channels in this circuit as the input was stepped through the four possible signal levels. The latter was carried out by synthesizing two of the signals using an input AOM and adding the output of the adder from the previous cycle, via the feedback loop and a beam-splitter, as the third.

This single channel demonstration could be extended to a parallel system in which each element of a logic array is acting as an independent adder unit. For example, addition of two four-bit numbers in parallel would require only five pixels, with five (shifted) interconnect channels for the carries. It is interesting to note that to implement the equivalent adder circuit using electronic NOR-gates would require 70 gates.

4.4.6 Communication Applications

(a) Optical Data Switching in a Fiber Link. With the widespread installation of fiber-optic communication systems, which transmit digital *optical* signals, there is considerable interest in using optical switching devices to preserve the advantages of the optical domain. In this case, it is the large-bandwidth capacity of optical interconnects that is important, rather than their potential for high parallelism. At present, fiber data-links are ultimately limited in throughput not by the fiber capacity but by the electronic/optoelectronic components used to transmit and receive the data. The development of fast all-optical or hybrid-optical switches, e.g. for high speed multiplexing or wideband routing, could permit considerable gains in the performance of future fiber networks.

The first experiment demonstrating the use of a bistable Fabry-Perot device in conjunction with a fiber-optic data transmission link was carried out in a Bell/Arizona collaboration in 1984 [4.61]. A cavity containing a nonlinear multiple-quantum-well GaAs/GaAlAs structure was used as a bistable switching element. It was placed at the center of a 2 km length of fiber such as to switch through a stream of (870 nm) 500 ns clock pulses (1.5 MHz) under the control of a counterpropagating stream of (835 nm) picosecond data pulses. The overall layout is summarized in Fig. 4.40. The effect of each logic-1 data pulse was to switch the nonlinear element, held within the bistable region by the clock pulse, into the highly transmitting state, hence gating the remainder of the clock pulse through to the output fiber. Because the latched switch only re-set on termination of the clock pulse, the duration of the output pulses could be varied by the timing of the short data pulses within the clock window.

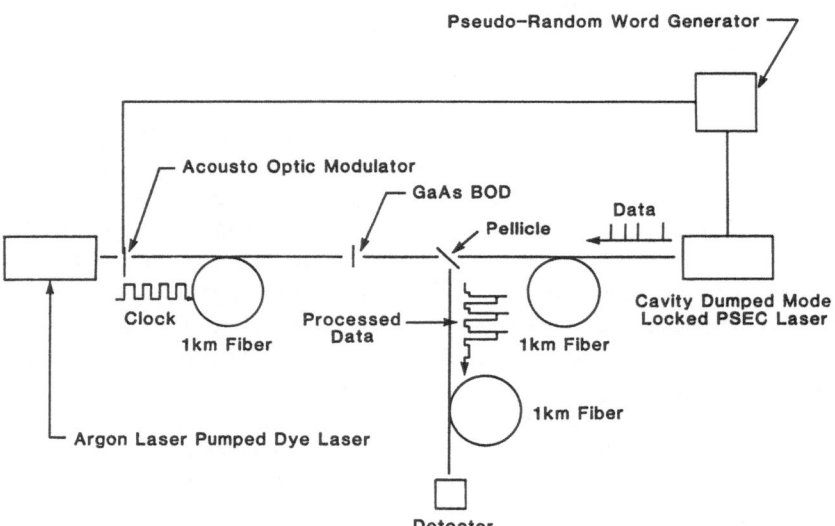

Fig. 4.40. Schematic of experimental fiber-optic system. The ∼ 1 MHz clock pulses were transmitted by the GaAs bistable optical device (BOD) only when a data pulse was simultaneously incident

The experiment also demonstrated the use of a signal of a different wavelength from the holding wavelength. This ability of bistable devices to transfer information from one wavelength to another was also demonstrated by *Seaton* et al. [4.62], using 1.06 μm pulses (once again picosecond duration) to switch a cw 5.5 μm wavelength beam using an InSb etalon. More recently, *Wherrett* et al. [4.63] have switched a ZnSe NLIF, operating cw at 633 nm, with picosecond 532 nm pulses.

(b) Spatial Switching with Bistable Etalons. Optical fiber systems need spatial switches capable of routing signals from an input fiber to one of a number of output fibers. Such switches would ideally be data transparent; i.e. acting as an optical relay that may not necessarily respond very fast but nonetheless has a wide throughput bandwidth. One such device currently being studied is the electro-optic directional coupler, which relies on interference effects between two coupled waveguides. The Fabry-Perot interferometer may be regarded as an analogous device, switching between reflecting and transmitting states as a result of small changes induced in the cavity thickness. By using *nonlinear* Fabry-Perot devices, switching can be controlled optically and, in addition, if they are bistable, their latching response can be usefully exploited.

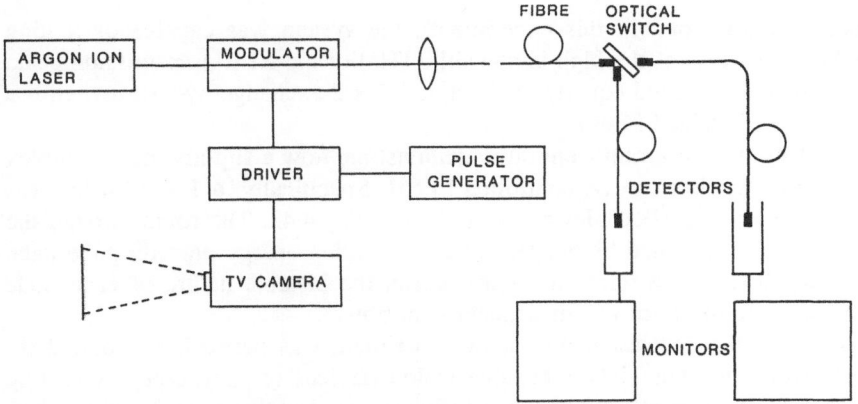

Fig. 4.41. Schematic of experimental layout used to demonstrate all-optical signal routing between fibers: a 1 × 2 switch. When the ZnSe NLIF (optical switch) was in its "off" state, the video data was directed to the reflection output, and when "on", to the transmission output. The data, holding power, and control pulses (used to switch between states of the device) were all incident via the input fiber

Figure 4.41 shows the layout of an experiment conducted by the Heriot-Watt group to demonstrate this concept [4.64]. A 1 × 2 switch was constructed using a 514 nm ZnSe NLIF. Optical power (hold beam), control signals and data were all supplied via an input fiber. No electrical connections were required in the vicinity of the switch. By switching the device with a control pulse (distinguished by its greater energy compared to the data pulses), the optical information could be routed to either one of the two output fibers. Despite a relatively slow device

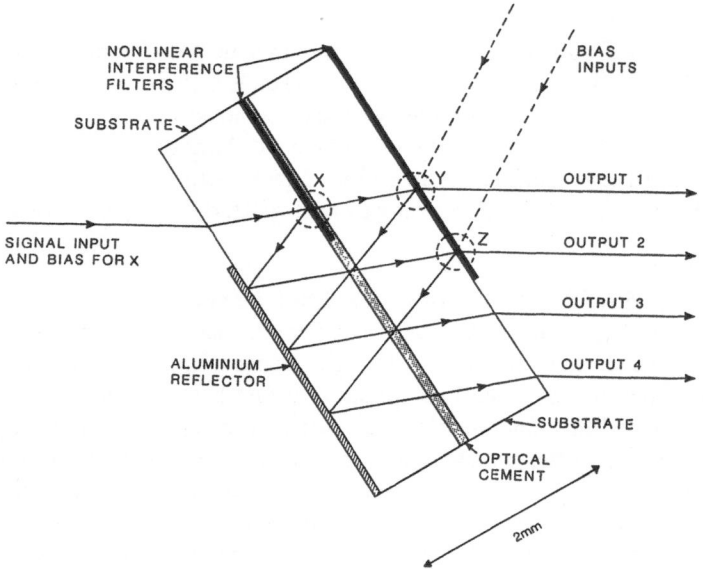

Fig. 4.42. An experimental 1 × 4 all-optical routing switch, based on ZnSe NLIFs

response (millisecond in this experiment), the system was capable of routing video data because of the high bandwidth (THz) of the NLIF transmission peak. This same device could equally well act as a 2 × 2 exchange/by-pass switch if a second input is added [4.64].

The Heriot-Watt group went on to demonstrate how a slightly more complex set of such switches can be controlled [4.65]. Specifically, a 1 × 4 switch was constructed using 3 NLIF devices as shown in Fig. 4.42. The route through the network was determined by header pulses of varying energy preceding the data. Once switched by the relevant control pulse, the bistable nature of each node maintained the route for the subsequent data flow.

Although the contrast in the latter experiment was particularly poor, these studies show the potential of bistable etalon devices (of any type) as routing switches. In practice, it would appear difficult to exploit this specific approach in large telecommunications networks because of insertion loss limitations as well as the contrast problem; however, in small-scale fiber systems, particularly where environmental factors make the distribution of electrical power dangerous, such all-optical networks could be extremely valuable.

(c) Compare-and-Exchange Switches. A disadvantage of the coding scheme for route setting that was described in the previous section is its partly analog nature: each level in the network responded to control pulses of less and less energy. This factor clearly limits how much the network could be extended. A preferable approach is one that relies on digital binary headers. This could be implemented in the case of the latter networks by changing from the asynchronous to a synchronous (clocked) switching approach [4.66].

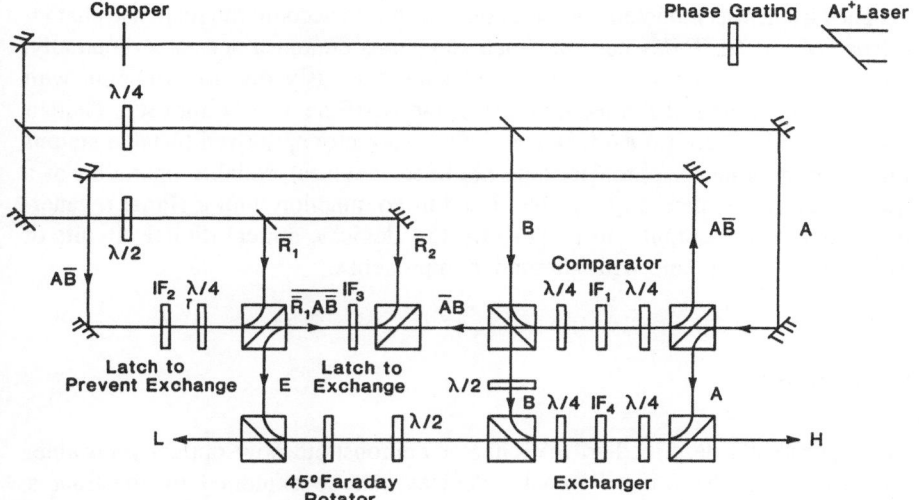

Fig. 4.43. Schematic of an experimental layout for an all-optical compare-and-exchange signal routing module. IF: ZnS NLIF; $\lambda/2, \lambda/4$: half-, quarter-wave plates. A,B represent two multi-bit binary numbers and H,L represent the outputs to which the higher and lower numbers (and the subsequent data) are directed

An alternative method is to include an optical-logic unit that decodes the header words and sets the routing appropriately. One example of this is a system with a unit that compares two binary numbers and sets the state of a 2×2 spatial switch according to which is larger. Such a system has been designed and investigated experimentally in a collaboration between the Arizona group and the BDM Corporation [4.67]. Figure 4.43 shows a schematic of the complete system. It consisted of four ZnS NLIF devices. The first of these, providing simultaneously the logic functions "a AND NOT b" and "b AND NOT a", acted as a comparator with its two outputs indicating which of inputs a and b was greater. The second two NLIFs were latching elements that were set by the comparator according to the first instance of a difference between the two inputs (the most significant bits headed the control word). The fourth NLIF element was a 2×2 spatial switch, of the type described in the previous section, controlled by the two latching elements. In this case, it was operated in an active mode, such that when it was required to be in its transmitting state it switched on each data pulse. As pointed out in the description of the system [4.67], this element could also be operated in the (wideband) data transparent mode, as demonstrated previously in the Heriot-Watt work [4.64].

A normal incidence geometry was used throughout the circuit, permitting optimization of switch powers (Sect. 4.3.2, [4.19]) and ensuring compatibility with possible operation, using switch arrays on each NLIF, of parallel channels through the system. Incident and back-reflected signals were separated in the usual manner by including 1/4-wave plates and polarizing beam-splitters.

The circuit was modeled numerically, taking into account the response characteristics of the NLIF devices, and each stage was demonstrated experimentally. The complete system was not operated simultaneously due to problems with switch power drift and limited contrast of the NLIF devices being used. Neither of these problems are fundamental, and, with correctly optimized bistable etalons (either optothermal or optoelectronic nonlinear devices), reliable operation of a system like this is perfectly feasible. Used in conjunction with a signal regeneration stage at the output and sufficiently fast devices, optical digital circuits of this type could compete with alternative approaches.

4.5 Conclusions

The experiments described in this chapter demonstrate how optically bistable, nonlinear Fabry-Perot interferometer devices can be exploited to construct a range of digital optical systems. Such systems correspond to the required building blocks for more extensive digital optical processors or optical switching computers. The successful demonstration of a full range of binary logic functions, logic level restoration, signal fan-out/fan-in and parallel operations has shown that, at least in principle, a complete optical computer could be constructed from devices of this type using the techniques described. What remains to be determined is whether such a system can (i) be engineered in a practical fashion, i.e. compact, mechanically stable, with good noise tolerance, acceptable power consumption, etc.; (ii) outperform by a significant factor the processing capabilities of alternative technologies, and (iii) prove itself competitive on economic grounds.

One crucial component that plays a major part in determining the future prospects of digital optical systems is the logic plane. This chapter has been intentionally limited to experiments based on bistable Fabry-Perot devices, and these were mainly, because of their simplicity and operational convenience, nonlinear interference filters (NLIFs). Despite the advantages of NLIFs over currently available alternatives, *all* the demonstration circuits were limited by the restricted performance characteristics of the logic elements. Nonetheless, by building such systems, valuable information has been obtained regarding the feasibility of these concepts, the required architectures and operating characteristics of all-optical processing systems, and the required device specifications.

Three approaches to the development of ideal logic plane devices can be identified at this stage. (i) The limits of NLIF devices remain to be explored. Reliable, stable NLIFs with $1–10\,\mu s$ switch times requiring $100\,\mu W–1\,mW$ of optical power are currently under development. Switch powers as low as $14\,\mu W$ have already been demonstrated (to date at millisecond time scales) by using liquid crystal spacers in Fabry-Perot cavities [4.68]. Wide availability of such improved NLIF devices will permit the extension of the limited demonstrations described here to complete working systems (e.g. see Sect. 4.4.4d–e). (ii) Further development of hybrid optical logic elements such as the liquid-crystal optically addressed SLMs or the GaAs-MQW SEED is proceeding. These devices have also been used in several optical circuit experiments not described here (e.g. see [4.69]). The liquid crystal SLMs have the advantage of high sensitivity but, at least at present, are slow compared to NLIF devices. Future hybrid SLMs exploiting ferroelectric liquid crystals may overcome this limitation. The SEEDs, with their low switch energies ($<$ 1 pJ), look an interesting prospect, especially the symmetric-SEED structure with its compatibility with differential logic schemes [4.70]. However, at present they are very much state-of-the-art semiconductor devices fabricated by high precision MBE growth techniques and could prove difficult to manufacture commercially in large arrays. (iii) Nonlinear (optoelectronic) GaAs-based Fabry-Perot structures, grown as single-crystal interference filters also look promising, provided carrier lifetimes can be controlled properly. Such bistable devices could have $\sim$ 100 ns response time and $<$ 1 mW power requirement if nonradiative carrier recombination processes could be sufficiently inhibited (e.g. by reductions in flaw densities). Advances in techniques for heat-sinking such elements so as to permit continuous operation [4.24, 71] are increasing the practical potential of these devices.

In addition to the development of optimized logic planes, other technologies need to be advanced. High efficiency optical interconnect systems are required, most probably based on volume hologram devices. Dynamically reconfigurable interconnects remain a desirable target but may not be essential if well–designed fixed interconnect architectures are achieved. In the absence of other dominant factors, the availability of efficient laser sources may determine the optimum operating wavelength. There is also a need for longer-term (nonvolatile) optical image storage with parallel read/write capability for rapid data transfer. Finally, there remains a need for algorithm development, in particular in the area of non-local interconnections (such as the perfect shuffle), for which the advantages of optics over electronics are strongest. Overall, it is clear that it will be the demonstration systems in which all these components are brought together that will highlight the specifications required, both of these components and of processing architectures. Such investigations will continue to play a key role in driving this research field forward to a range of realistic optical processing applications.

References

4.1   S.F. Reddaway: *High Speed Computation*, Proc. NATO Workshop (Springer, Berlin, Heidelberg 1983)
       J.B.G. Roberts, P. Simpson, B.C. Merrifield, J.F. Cross: IEEE Proc. **131**, 603 (1984)
4.2   W.D. Hills: *The Connection Machine* (MIT, Cambridge, MA 1985) p. 190
4.3   K. Batcher: IEEE Trans. Comp. **29**, 837 (1980)
4.4   M. Duff, T. Fountain: *Cellular Logic Image Processing* (Academic, New York 1986)
4.5   L.A. Zadeh: Inf. Control. **8**, 338 (1969)
4.6   B.S. Wherrett: Appl. Opt. **24**, 2876 (1985)
4.7   A.C. Walker: Appl. Opt. **25**, 1578 (1986)
4.8   H.M. Gibbs: *Optical Bistability: Controlling Light with Light* (Academic, New York 1985)

4.9   H.M. Gibbs, S.L. McCall, T.N.C. Venkatesan: Phys. Rev. Lett. **36**, 1135 (1976)
4.10   H.M. Gibbs, S.L. McCall, T.N.C. Venkatesan, A.C. Gossard, A. Passner, W. Wiegmann: Appl. Phys. Lett. **35**, 6 (1979)
4.11   D.A.B. Miller, S.D. Smith, A.M. Johnston: Appl. Phys. Lett. **35**, 658 (1979)
4.12   D.A.B. Miller, C.T. Seaton, M.E. Prise, S.D. Smith: Phys. Rev. Lett. **47**, 197 (1981)
4.13   B.S. Wherrett: Proc. R. Soc. Lond. A **390**, 373 (1983)
4.14   F.A.P. Tooley: Appl. Opt. **26**, 1741 (1987)
4.15   E. Garmire, J.H. Marburger, S.D. Allen, H.G. Winful: Appl. Phys. Lett. **34**, 374 (1979)
4.16   F.A.P. Tooley, S.D. Smith, C.T. Seaton: Appl. Phys. Lett. **43**, 807 (1983)
4.17   B.S. Wherrett: IEEE J. QE-**20**, 645 (1984)
4.18   J.W. Goodman: Opt. Acta **32**, 1489 (1985)
4.19   G.S. Buller, C.R. Paton, S.D. Smith, A.C. Walker: Opt. Commun. **70**, 522 (1989)
4.20   S.D. Smith, A.C. Walker, B.S. Wherrett, F.A.P. Tooley, J.G.H. Mathew, M.R. Taghizadeh, I. Janossy: Appl. Opt. **25**, 1586 (1986)
4.21   R. Cush: Electron. Lett. **23**, 619 (1987)
4.22   B.S. Wherrett, J.F. Snowdon: *Optical Computing 88*, ed. by J.W. Goodman, P. Chavel, G. Roblin; Proc. SPIE **963**, 15 (1989)
4.23   O. Sahlen, V. Olin, E. Masseboeuf: Appl. Phys. Lett. **50**, 1559 (1987)
4.24   R. Kuszelewicz, J.L. Oudar, R. Azoulay, J.C. Michel, J. Brandon: J. de Phys. Colloq. Suppl. **49**, C2–193 (1988)
4.25   C.R. Pidgeon, S.D. Smith: J. Opt. Soc. Am. **54**, 1459 (1964)
4.26   F.V. Karpushko, G.V. Sinitsyn: J. Appl. Spectrosc. (USSR) **29**, 1323 (1978)
4.27   D.A. Weinberger, H.M. Gibbs, C.F. Li, M.C. Rushford: J. Opt. Soc. Am. **72**, 1769 (1982)
4.28   S.D. Smith, J.G.H. Mathew, M.R. Taghizadeh, A.C. Walker, B.S. Wherrett, A. Hendry: Opt. Commun. **51**, 357 (1984)
4.29   R.J. Campbell, J.G.H. Mathew, S.D. Smith, A.C. Walker: J. Mod. Opt. **36**, 323 (1989)
4.30   Y.T. Chow, B.S. Wherrett, E. Van Stryland, B.T. McGuckin, D. Hutchings, J.G.H. Mathew, A. Miller, K. Lewis: J. Opt. Soc. Am. B **3**, 1535 (1986)
4.31   G.R. Olbright, N. Peyghambarian, H.M. Gibbs, H.A. MacLeod, F. Van Milligen: Appl. Phys. Lett. **45**, 1031 (1984)
4.32   S.P. Apanasevich, F.V. Karpushko, G.V. Sinitsyn: Sov. J. QE-**14**, 873 (1984)
4.33   I. Janossy, J.G.H. Mathew, E. Abraham, M.R. Taghizadeh, S.D. Smith: IEEE J. QE-**22**, 2224 (1986)
4.34   A.C. Walker, M.R. Taghizadeh, J.G.H. Mathew, I. Redmond, R.J. Campbell, S.D. Smith, J. Dempsey, G. Lebreton: Opt. Eng. **27**, 38 (1988)
4.35   I. Janossy, M.R. Taghizadeh, J.G.H. Mathew, S.D. Smith: IEEE J. QE-**21**, 1447 (1985)
4.36   B.S. Wherrett, D. Hutchings, D. Russell: J. Opt. Soc. Am. B **3**, 351 (1986)
4.37   E. Abraham, A.K. Kar, M.R. Suttie, R.M. Harris, A.C. Walker, S.D. Smith: J. Appl. Phys. **64**, 3393 (1988)
4.38   J.L. Jewell, A. Scherer, S.L. McCall, A.C. Gossard, J.H. English: Appl. Phys. Lett. **51**, 94 (1987)
4.39   A.C. Walker: Opt. Commun. **59**, 145 (1986)
4.40   R.J. Keyes: Opt. Acta **32**, 525 (1985)
4.41   S.D. Smith, F.A.P. Tooley: *Optical Bistability 2*, ed. by C.M. Bowden, C. Ciftan, H. Robl (Plenum, New York 1981) p. 215
4.42   A.C. Walker, F.A.P. Tooley, M.E. Prise, J.G.H. Mathew, A.K. Kar, M.R. Taghizadeh, S.D. Smith: Philos. Trans. R. Soc. London A **313**, 249 (1984)
4.43   S.D. Smith, I. Janossy, H.A. MacKenzie, J.G.H. Mathew, J.J. Reid, M.R. Taghizadeh, F.A.P. Tooley, A.C. Walker: Opt. Eng. **24**, 569 (1985)
4.44   R.G.A. Craig, G.S. Buller, F.A.P. Tooley, S.D. Smith, A.C. Walker, B.S. Wherrett: Optical Computing, Salt Lake City 1989, Postdeadline Paper. Also: accepted for publication in Appl. Optics (1990)
4.45   A.C. Walker, B.S. Wherrett, S.D. Smith: *Nonlinear Optical Materials II*; SPIE Proc. **1127**, 17 (1989)
4.46   S.S. Tarng, K. Tai, J.L. Jewell, H.M. Gibbs, A.C. Gossard, S.L. McCall, A. Passner, T.N.C. Venkatesan, W. Wiegmann: Appl. Phys. Lett. **40**, 205 (1982)
4.47   S.D. Smith, A.C. Walker, F.A.P. Tooley, J.G.H. Mathew, M.R. Taghizadeh: *Optical Bistability III*, ed. by H.M. Gibbs, P. Mandel, N. Peyghambarian, S.D. Smith, Springer Proc. Phys., Vol. 8 (Springer, Berlin, Heidelberg 1986) p. 8
4.48   S.D. Smith, A.C. Walker, F.A.P. Tooley, B.S. Wherrett: Nature **325**, 27 (1987)

4.49   S.D. Smith, A.C. Walker, B.S. Wherrett, F.A.P. Tooley, N. Craft, J.G.H. Mathew, M.R. Taghizadeh, I. Redmond, R.J. Campbell: Opt. Eng. **26**, 45 (1987)

4.50   M.T. Tsao, L. Wang, R. Jin, R.W. Sprague, G. Gigioli, H.M. Kulcke, Y.D. Li, H.M. Chou, H.M. Gibbs, N. Peyghambarian: Opt. Eng. **26**, 41 (1987)

4.51   L. Wang, V. Esch, R. Feinleib, L. Zhang, R. Jin, H.M. Chou, R.W. Sprague, H.A. MacLeod, G. Khitrova, H.M. Gibbs, K. Wagner, D. Psaltis: Appl. Opt. **27**, 1715 (1988)

4.52   A. Huang: Proc. IEEE 10th Int. Optical Computing Conf., No. 83CH1880-4 (1983) p. 13

4.53   M.R. Taghizadeh, I. Redmond, B. Robertson, A.C. Walker, S.D. Smith: *Holographic Optics II*; SPIE Proc. **1136**, 265 (1989)

4.54   B.S. Wherrett: SPIE Proc. **769**, 7 (1987)

4.55   G. Khitrova, L. Wang, V. Esch, R. Feinleib, H.M. Chou, R.W. Sprague, H.A. MacLeod, H.M. Gibbs, K. Wagner: *Optical Computing and Nonlinear Materials*; SPIE Proc. **881**, 60 (1988)

4.56   Y. Owechko, G.J. Dunning, E. Marom, B.H. Soffer: Appl. Opt. **26**, 1900 (1987)

4.57   K.H. Brenner, A. Huang: *Optical Computing*, Technical Digest (OSA, Washington, DC 1985), paper WA4

4.58   L. Wang, H.M. Chou, H.M. Gibbs, G.C. Gigioli, G. Khitrova, H.M. Kulcke, R. Jin, H.A. MacLeod, N. Peyghambarian, R.W. Sprague, M.T. Tsao: *Digital Optical Computing*, SPIE Proc. **752**, 14 (1987)

4.59   B.S. Wherrett: Opt. Commun. **56**, 87 (1985)

4.60   F.A.P. Tooley, N.C. Craft, S.D. Smith, B.S. Wherrett: Opt. Commun. **63**, 365 (1987)

4.61   T. Venkatesan, P.J. Lemaire, B. Wilkens, L. Soto, A.C. Gossard, W. Wiegmann, J.L. Jewell, H.M. Gibbs, S.S. Tarng: Opt. Lett. **9**, 297 (1984)

4.62   C.T. Seaton, S.D. Smith, F.A.P. Tooley, M.E. Prise, M.R. Taghizadeh: Appl. Phys. Lett. **42**, 131 (1983)

4.63   B.S. Wherrett, Y.T. Chow, A.K. Rhoomy-Darzi, A.D. Lloyd: Phys. Scr. **T25**, 247 (1989)

4.64   C.R. Paton, S.D. Smith, A.C. Walker: *Photonic Switching*, ed. by T.K. Gustafson, P.W. Smith, Springer Ser. Electron. Photon., Vol. 25 (Springer, Berlin, Heidelberg 1988) p. 59

4.65   G.S. Buller, C.R. Paton, S.D. Smith, A.C. Walker: Appl. Phys. Lett. **53**, 2465 (1988)

4.66   G.S. Buller: PhD Thesis, Heriot-Watt University, Edinburgh

4.67   L. Zhang, R. Jin, C.W. Stirk, G. Khitrova, R.A. Athale, H.M. Gibbs, H.M. Chou, R.W. Sprague, H.A. MacLeod: IEEE J. SAC-6, 1273 (1988)

4.68   A.D. Lloyd, B.S. Wherrett: J. de Phys. Colloq. Suppl. **49**, C2–141 (1988)

4.69   A.A. Sawchuk, T.C. Strand: Proc. IEEE **72**, 758 (1984)

4.70   A.L. Lentine, H.S. Hinton, D.A.B. Miller, J.E. Henry, J.E. Cunningham, L.M.F. Chirovsky: Appl. Phys. Lett. **52**, 1419 (1988)

4.71   E. Masseboeuf, O. Sahlen, U. Olin, N. Nordell, M. Rask, and G. Landgren: Appl. Phys. Lett. **54**, 2290 (1989)

5. Optical Processing with Nonlinear Photorefractive Crystals

H. Rajbenbach, J.-P. Huignard and P. Günter

With 29 Figures

The development of real-time devices for the manipulation of two-dimensional time-varying coherent wavefronts is of major interest for optical processing applications. It was recognized early on that the photoinduced index change in electro-optic crystals (the photorefractive effect) provides an opportunity for implementing highly nonlinear functions by means of low-power cw lasers. A wide variety of operations have already been performed, including image amplification and correlation, phase conjugation, binary two-dimensional logic, spatial light modulation, associative recall and beam-steering besides numerous media display photorefractive properties at UV, visible, or infrared wavelengths. Current photorefractive materials consist of electro-optic oxides such as $LiNbO_3$, $BaTiO_3$, $KNbO_3$, $Bi_{12}SiO_{20}$ (or BSO), $Bi_{12}GeO_{20}$ (or BGO), $Ba_{1-x}Sr_xNb_2O_6$ (or SBN), and semiconductors such as GaAs, InP and CdTe. This chapter reviews theoretical and experimental work on the photorefractive effect with regard to applications in the field of optical processing. The following section presents the mechanisms of photorefractive grating recording with emphasis on the relevant parameters for practical applications. In Sect. 5.3, the conditions for efficient wave mixing in photorefractive media are analyzed. Various applications, considered in the context of optical computing, are described in the final sections.

5.1 Characteristic Parameters of Photorefractive Crystals

The light induced changes of refractive indices in electro-optic crystals are based on the spatial modulation of charges by nonuniform illumination (Fig. 5.1). Electrons (or holes) are photoexcited from impurity centers present in the material and, upon migration, are retrapped at other locations, leaving behind positive or negative charges of ionized trap centers [5.1–3]. The photoexcited charges will be re-excited and retrapped until they finally drift out of the illumination region. The resulting space-charge field between the ionized donor centers and the trapped charges modulates the refractive indices via the electro-optic effect. Uniform illumination erases the space charge fields and brings the crystal back to its original state (optical erasure).

The complete mathematical description of grating formation in photorefractive crystals has been derived by *Kukhtarev* et al. [5.4]. From this model, the photoinduced index modulation at the saturation region is given by

Nonlinear Photonics Editors: H.M. Gibbs · G. Khitrova · N. Peyghambarian
© Springer-Verlag Berlin Heidelberg 1990

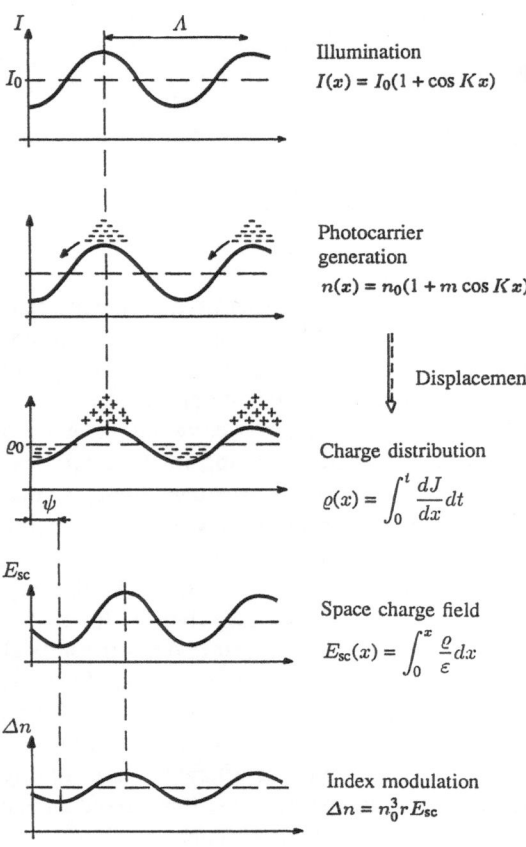

Illumination
$I(x) = I_0(1 + \cos Kx)$

Photocarrier
generation
$n(x) = n_0(1 + m \cos Kx)$

Displacement

Charge distribution
$\varrho(x) = \int_0^t \frac{dJ}{dx}dt$

Space charge field
$E_{sc}(x) = \int_0^x \frac{\varrho}{\varepsilon}dx$

Index modulation
$\Delta n = n_0^3 r E_{sc}$

Fig. 5.1. Mechanism of photorefractive grating recording in electro-optic crystals: light with a spatially periodic intensity (I) resulting from the interference of two beams in the material rearranges the charge density (ϱ), which causes a periodic field (E_{sc}) according to Poisson's equation. This electric field then causes a change in the refractive index of the crystal (Δn) by the linear electro-optic effect

$$\Delta n = 2n_0^3 r \frac{\beta^{1/2}}{1 + \beta} \left(\frac{E_0^2 + E_D^2}{(1 + E_D/E_q)^2 + (E_0^2/E_q^2)} \right)^{1/2} \tag{5.1}$$

and the spatial shift ψ between the incident fringe pattern and the photoinduced index modulation is given by

$$\tan \psi = \frac{E_D}{E_0} \left(1 + \frac{E_D}{E_q} + \frac{E_0^2}{E_D E_q} \right) , \tag{5.2}$$

where r is the electro-optic coefficient, n_0 is the background refractive index of the medium, β is the incident intensity ratio of the two interfering beams, E_0 is an externally applied field, E_D is the diffusion field, and E_q is the maximum field which would correspond to a complete separation of the positive and negative charges by one grating period. The expressions for these fields are

$$E_D = 2\pi k_B T/e\Lambda ; \quad E_q = eN_A\Lambda/2\pi\varepsilon_0\varepsilon ,$$

where N_A is the trap density in the crystal volume, Λ is the fringe spacing, ε is the relative static dielectric constant, T is the temperature, e is the electron charge, and ε_0 is the free space permittivity.

In this section, we will briefly discuss some of the most important parameters involved in the applications of photorefractive crystals to optical processing i.e., the crystal sensitivity [5.5, 6], the grating buildup time constants and the steady-state diffraction efficiency.

5.1.1 Crystal Sensitivity

The photorefractive sensitivity is defined as the refractive index change Δn per unit absorbed energy density:

$$S = \Delta n / \alpha I_0 \tau \,,$$

where α is the crystal absorption coefficient at recording wavelength λ, τ is the crystal response time, and I_0 the incident power density. This definition is a useful figure of merit for comparing materials having different absorption coefficients at a given wavelength [5.7]. The response time of a photorefractive crystal is the dielectric relaxation time multiplied by a function of different parameters such as applied field E_0, grating spacing Λ, drift, and diffusion lengths (respectively r_E and r_D) of the photocarriers:

$$S = \frac{1}{2} n_0^3 \frac{r}{\varepsilon \varepsilon_0} F(E_0, \lambda, r_E, r_D) \,. \tag{5.3}$$

Since n_0 and r/ε are nearly constant for all electro-optic inorganic crystals, the photorefractive sensitivity is mainly determined by the recording conditions and by the relative values of the drift and diffusion lengths compared to the grating spacing. The sensitivity S reaches a maximum value when the excited photocarriers drift or diffuse over distances equal to or larger than the grating spacing. The upper limit of S for an elementary grating ($\beta = 1$) and unit quantum efficiency is [5.7]

$$S_{\max} = n_0^3 r e \Lambda / 4 \pi \varepsilon \varepsilon_0 h v$$

This can be estimated to $0.1 \, \mathrm{cm}^3 \, \mathrm{J}^{-1}$ for $\lambda = 0.5 \, \mu$m. This optimum photorefractive sensitivity is reached in efficient photoconductive crystals such as KTN, BSO, BGO and GaAs.

Another figure of merit commonly used for experiments with photorefractive crystals is the energy per unit area W to write an elementary grating ($\beta = 1$), having 1% or a few percent efficiency in a crystal 1 mm (or a few millimeters) thick. This figure of merit enables a ready comparison of fast and slow materials illuminated with the same incident beam intensity I_0 [5.7]. Writing energies as low as $W \sim 100 \mu J \, \mathrm{cm}^{-2}$ are available in photorefractive crystals. It must be noted that these values of recording energies in dynamic materials are nearly equivalent to high resolution silver halide plates.

153

5.1.2 Steady-State Diffraction Efficiency

The diffraction efficiency η of a thick phase transmission grating with a peak-to-peak index modulation $2\Delta n$ is derived from the *Kogelnik* formula [5.8]

$$\eta = \exp\left(-\alpha d/\cos\theta\right)\sin^2(\pi d\Delta n/\lambda\cos\theta)$$

where θ is the Bragg angle inside the crystal and d is the thickness of the crystal. High values of the photoinduced index change and therefore of the diffraction efficiency are obtained in materials with high electro-optic coefficients, such as ferroelectric crystals $BaTiO_3$ [5.9], SBN [5.10] ($r \approx 10^3\,\text{pm/V}$) or $KNbO_3$ [5.11] ($r = 60\,\text{pm/V}$). In other materials having low electro-optic coefficients, such as BSO [5.12, 13], BGO, GaAs or InP [5.14, 15] ($r \approx 1-3\,\text{pm/V}$), Δn can be increased with an externally applied electric field E_0 until saturation occurs for $E_0 = E_q$. As an example, Fig. 5.2 shows the diffraction efficiency of a crystal of BGO as a function of the applied field [5.16]. Almost 100% efficiency is measured for an optimized fringe spacing of 20 μm.

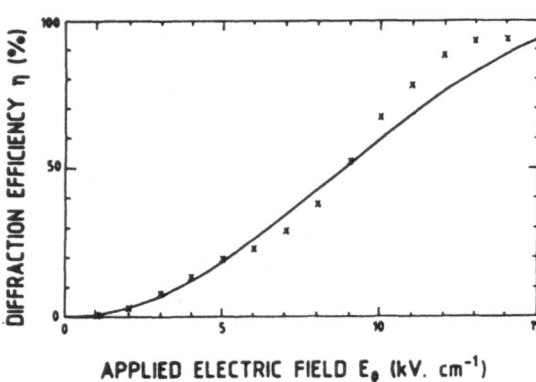

Fig. 5.2. High diffraction efficiencies are obtained with photorefractive volume gratings. In BGO, almost 100% is measured when a high external dc electric field is applied to the crystal (drift mode). Grating spacing $\Lambda = 20\,\mu\text{m}$; beam ratio $\beta = 1$; crystal thickness $d = 10\,\text{mm}$; wavelength $\lambda = 0.514\,\mu\text{m}$. (From [5.16])

5.1.3 Response Time of the Photorefractive Effect

The time constant for buildup of a grating is also an important characteristic of the photorefractive effect. The refractive index changes are due to electro-optic effects driven by space-charge fields, and the time required to record a grating depends on the efficiency of the charge generation and transport process. The inertia in the nonlinear response of photorefractive media constitutes an important difference from other nonlinear media where the refractive index change is of electronic origin and thus occurs instantaneously. Under continuous wave illumination, the crystal response time is [5.7]:

$$\tau = \tau_{di}\frac{(1+\tau_R/\tau_D)^2 + (\tau_R/\tau_E)^2}{[1+(\tau_R\tau_{di}/\tau_D\tau_I)](1+\tau_R/\tau_D) + (\tau_R/\tau_E)^2(\tau_{di}/\tau_I)}, \tag{5.4}$$

where τ_{di} is the dielectric relaxation time of the crystal:

$$\tau_{di} = \varepsilon\varepsilon_0/n_0\mu e \; ;$$

n_0 is the free carrier concentration due to the incident illumination I_0:

$$n_0 = \tau_R\alpha\phi I_0/h\upsilon \; ;$$

μ is the mobility of the photocarriers and ϕ the quantum efficiency. The charge recombination time τ_R may be written as

$$\tau_R = (\gamma_R N_A)^{-1} \; ,$$

where γ_R is the recombination coefficient, τ_E and τ_{di} are the drift and diffusion time of the charges, given respectively by

$$\tau_E = 1/K\mu E_0 \; , \quad \tau_{di} = 1/\mu k_B T K^2 \; .$$

$K = 2\pi/\Lambda$ and τ_I is the inverse of the sum of photogeneration rate sI_0 and ion recombination rate $\gamma_R n_0$:

$$\tau_I = (sI_0 + \gamma_R n_0)^{-1} \; .$$

A simple expression for the time dependence of the space-charge field during grating recording is

$$\Delta E_{sc} = m E_{sc}(1 - e^{-t/\tau}) \; .$$

During erasure by uniform illumination, the photoinduced space-charge field decreases according to

$$\Delta E_{sc} = m E_{sc}^{-t/\tau} \; ,$$

where E_{sc} is the initial amplitude of the field and m the incident modulation. The recording erasure cycle is therefore symmetrical as shown in Fig. 5.3 for photorefractive BSO crystals. The typical recording erasure time for an elementary grating of 10–100 ms corresponds to a cw incident intensity of 10–100 mW cm^{-2} at the blue or green line of the argon laser. BSO and GaAs are used as fast and sensitive materials, while crystals such as BaTiO$_3$ have large electro-optic coefficients but respond rather slowly, i.e., have a response time of a few seconds. Therefore, another important figure of merit for a photorefractive crystal will be the energy required to reach the steady-state diffraction efficiency; and this parameter often determines the crystal chosen for a particular application.

Photorefractive index gratings can be alternatively recorded and erased with short optical pulses [5.17, 18]. Although numerous studies have been performed in the nanosecond and picosecond regimes with LiNbO$_3$ [5.19], BaTiO$_3$ [5.20, 21], BSO [5.22], and GaAs [5.23, 24], most of the experimental research for parallel optical processing applications have involved low power cw sources, and the following sections will focus on this regime.

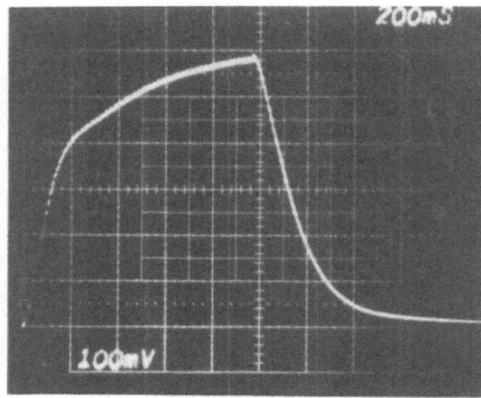

Fig. 5.3. Grating recording–erasing cycle in BSO at $\lambda = 0.568\,\mu$m. Incident beam intensity $120\,\text{mW cm}^{-2}$; $E_0 = 6\,\text{kV cm}^{-1}$; fringe spacing $\Lambda = 6\,\mu$m. Steady-state diffraction efficiency $\eta = 10\%$ monitored with a low power He-Ne laser. (From [5.25])

5.2 Beam Coupling in Photorefractive Crystals

The recording of phase volume gratings in photorefractive media leads to a stationary energy exchange between the two interfering beams. The resulting energy redistribution that has been observed in many electro-optic crystals (LiNbO$_3$, KNbO$_3$, BaTiO$_3$, BSO and GaAs) [5.3, 25, 26] is due to self-diffraction of the reference pump beam by the dynamic phase grating photoinduced in the crystal. More specifically, the self interference of the incident beam with the diffracted beam creates a new holographic grating which can add to (or subtract from) the initial one. Since the diffracted wave is phase delayed by $\pi/2$ with respect to the reading beam, the maximum energy transfer is obtained when the incident fringe pattern and the photoinduced index modulation are shifted by $\psi = \pi/2$ [5.1–4, 27, 28].

In photorefractive crystals, such a $\pi/2$ phase shift exists when the recording is by "diffusion" of photocarriers (no external applied electric field), as shown in Fig. 5.4. As a consequence, a permanent and efficient amplification of a low inten-

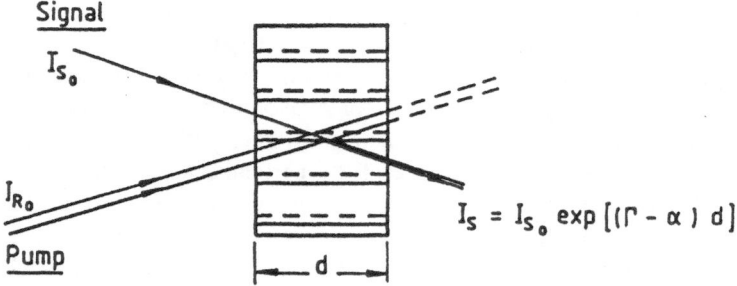

Fig. 5.4. Degenerate two-wave mixing (2WM) in photorefractive crystals. The continuous and dashed lines represent the maxima of the illumination and index pattern, respectively. In the diffusion mode (no applied field), the phase shift ψ is $\pi/2$ and large gains are observed in crystals such as BaTiO$_3$ or SBN

sity signal beam has been observed in crystals like LiNbO$_3$, BaTiO$_3$ or KNbO$_3$. If we now apply the coupled wave equations to the $\pi/2$ phase shifted component of the photoinduced index modulation, the coherent interaction between the two waves of respective amplitude R and S is described by [5.8]

$$\frac{dS}{dz} = \frac{1}{2}\Gamma\frac{R^2 S}{R^2 + S^2} - \frac{1}{2}\alpha R , \qquad \frac{dR}{dz} = \frac{1}{2}\Gamma\frac{RS^2}{R^2 S^2} - \frac{1}{2}\alpha S , \qquad (5.5)$$

and the transmitted signal beam intensity resulting from the dynamic two-beam coupling takes the form

$$I_S = I_{S0}\frac{\beta + 1}{\beta + \exp(\Gamma d)} \exp[(\Gamma - \alpha)d] , \qquad (5.6)$$

where Γ is the exponential gain coefficient of the interaction and is related to the maximum amplitude of the photoinduced index modulation Δn_S through

$$\Gamma = \frac{4\pi \Delta n_S}{\lambda \cos\theta} \sin\psi . \qquad (5.7)$$

ψ is the spatial phase shift of the grating and, in agreement with the previous arguments, Γ is maximum when $\psi = \pi/2$. In the case of a negligible pump depletion, the transmitted signal beam intensity is simply

$$I_S = I_{S0} \exp[(\Gamma - \alpha)d] , \qquad (5.8)$$

and therefore when the condition $\Gamma > \alpha$ is fulfilled, the incident signal exhibits gain and the photorefractive crystal may be regarded as a parametric amplifier.

A practical parameter for characterizing the energy transfer due to the two-beam coupling is the effective gain γ_0, defined by the ratio [5.28]

$$\gamma_0 = \frac{I_S \text{ with pump beam}}{I_S \text{ without pump beam}} . \qquad (5.9)$$

For the undepleted pump approximation we have

$$\gamma_0 = \exp(\Gamma d) .$$

Therefore, from the measurement γ_0, the value of the exponential gain coefficient Γ of the interaction can be easily deduced. Large values of Γ ($\Gamma = 20$ cm^{-1}) may be obtained in materials having large Δn_S when recording by diffusion ($\psi = \pi/2$), and this is the case for BaTiO$_3$, SBN, LiNbO$_3$ and KNbO$_3$. However, the same 2WM experiment performed with highly photoconductive BSO or GaAs crystals leads to very low beam coupling ($\Gamma = 1$ cm^{-1}) for the following reasons: (i) For diffusion, the required phase shift $\psi = \pi/2$ is established, but the steady-state index modulation is low and (ii) for drift, with an electric field applied to the crystal, the index modulation is much higher, but the corresponding phase shift ψ is negligible. However, efficient beam coupling can be obtained if the fringe

157

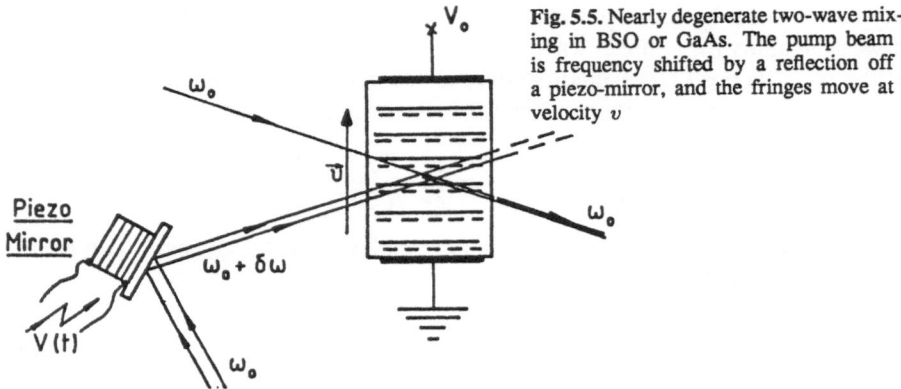

Fig. 5.5. Nearly degenerate two-wave mixing in BSO or GaAs. The pump beam is frequency shifted by a reflection off a piezo-mirror, and the fringes move at velocity v

pattern (or the crystal) is moved at a constant velocity. The speed is adjusted such that the index modulation is recorded at all times, but with a spatial phase shift with respect to the interference fringes. Clearly, the optimum fringe velocity will depend on the recording time constant τ, and when the interference pattern moving at velocity v is introduced into the coupled wave equations, the resulting gain coefficient Γ is [5.29]

$$\Gamma = \frac{4\pi \Delta n_S}{\lambda \cos \theta} \frac{Kv\tau}{(1 + K^2 v^2 \tau^2)} \; . \tag{5.10}$$

In a 2WM configuration such as that shown in Fig. 5.5, the fringe displacement increases the amplitude of the $\pi/2$ phase shifted component of the index modulation, and, consequently, efficient energy transfer is obtained in photorefractive

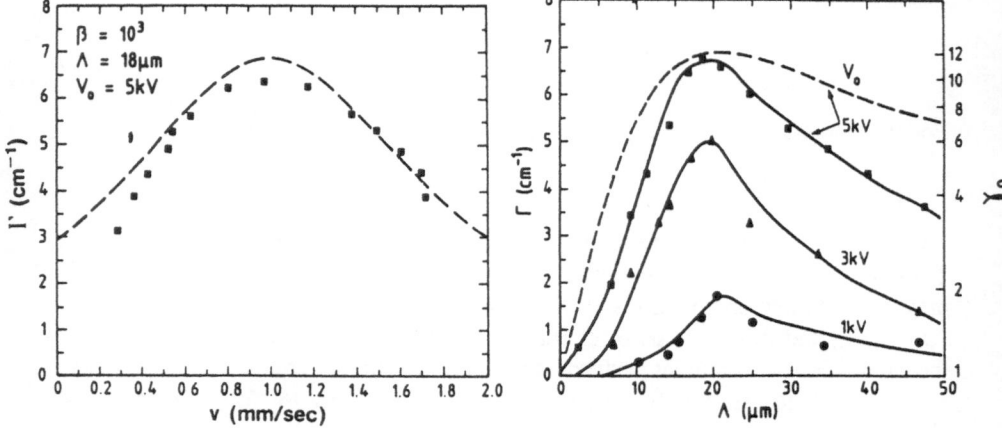

Fig. 5.6a,b. Evolution of the gain coefficient in GaAs at $\lambda = 1.06 \, \mu$m. (a) Large gains occur when the fringes move at the optimum velocity: the space-charge field is optimized and $\pi/2$ shifted with the illumination. (b) Exponential gain coefficient as a function of the grating fringe spacing for different applied voltages. $\beta = 10^3$. *Dashed lines*: theoretical plots. (From [5.26])

crystals like BSO [5.25] and GaAs [5.26]. The expected optimum fringe velocity is $v_0 = \Lambda(2\pi\tau)^{-1}$ which corresponds to a frequency detuning by $\delta\omega = \tau^{-1}$ of the reference beam [5.30]. This interaction is termed *nearly degenerate two-wave mixing*. A simple method of frequency detuning the reference beam by $\delta\omega$ is to use a piezo-mirror driven by a saw-tooth voltage [5.31]. Figure 5.6a shows the resonant dependence of the gain on the fringe velocity in GaAs.

Large photorefractive gains are also available when operating the crystal under alternating electric fields. This "nonstationary" interaction was first proposed by *Stepanov* et al. for improving the gain coefficient of $Bi_{12}TiO_{20}$ crystals in a degenerate interaction [5.32] and was also used for BSO [5.33] and semiconductor GaAs and InP [5.34, 35]. The equation describing the coupling mechanisms predict optimized gains equal to those of the moving grating techniques.

5.2.1 Influence of the Recording Parameters (Spatial Frequency, Fringe Velocity, Beam Ratio)

A precise knowledge of the spatial frequency response of photorefractive crystals is important for applications to coherent image amplification and optical signal processing.

Figure 5.6b shows the dependence of the GaAs nearly-degenerate 2WM gain Γ on the fringe spacing Λ and as a function of the applied voltage V_0. For each measurement, the fringe velocity is adjusted such that the maximum gain is obtained. The incident pump beam intensity is 40 mW cm^{-2} at the recording

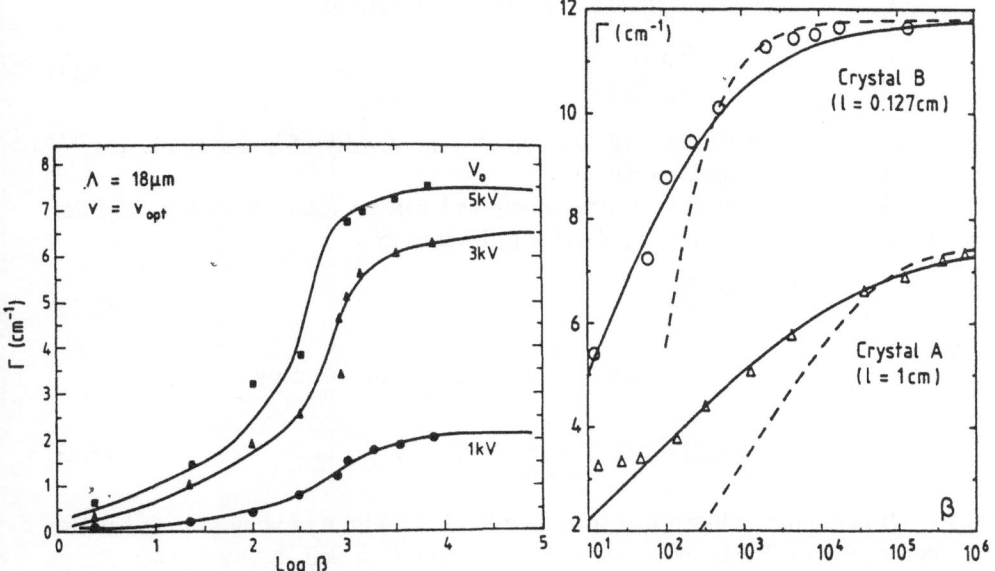

Fig. 5.7. Dependence of the gain coefficient in GaAs at $\lambda = 1.06\,\mu$m (*left*) and in BSO at $\lambda = 0.568\,\mu$m (*right*) on the incident beam ratio $\beta = I_{R0}/I_{S0}$. The nonlinear gain is explained by introducing the second-order term into the expansion of the space-charge fields. (From [5.25, 26])

wavelength $\lambda = 1.06 \, \mu m$ and the incident beam ratio is $\beta = 10^3$ (corresponding to a time constant $\tau = 120 \, ms$). These curves show a strong increase in the gain for $\Lambda \sim 20 \, \mu m$. Figure 5.7 represents the variation of Γ as a function of the incident beam ratio β.

The following points summarize the main conclusions that can be drawn from these curves: (i) High gain is available in photorefractive semiconductor GaAs when recording with a high electric field and moving the fringes at the optimum velocity. (ii) The gain of the amplifier is strongly dependent on the grating spatial frequency. (iii) The gain reaches saturation at high beam ratio. Consequently, a wide range of experimental conditions allow one to obtain a value of Γ in excess of the crystal absorption losses ($\alpha \approx 1.6 \, cm^{-1}$ at $\lambda = 1.06 \, \mu m$ for GaAs) and $\Gamma \approx 6-7 \, cm^{-1}$ for optimized recording conditions.

The dependence of Γ on the grating spatial frequency Λ^{-1} and the incident beam ratio β are described by *Kukhtarev*'s equations. Indeed, fitting the experimental plots with the theoretical predictions yields a set of numerical values of crystal parameters such as the mobility and the recombination coefficients. The starting point is the set of differential equations that describe the charge transport and trapping and for which the incident fringe illumination is

$$I(x, t) = I_0[1 + m \cos K(x - vt)] .$$

Results of *Valley* [5.36] and *Refregier* et al. [5.37] indicate the important features related to the fringe displacement in crystals such as GaAs and BSO. Derived from these references, the velocity that maximizes the imaginary part of the space charge field responsible for energy transfer is

$$v_{opt} = s(I_{R0} + I_{S0}) \frac{N_D}{N_A} \frac{E_q}{E_0} \frac{\Lambda}{2\pi} , \tag{5.11}$$

where s is the ionization cross section, N_D the density of donor atoms and N_A the density of acceptor atoms.

An optimum spacing grating exists and can be found from the condition $E_M E_q = E_0^2$ where $E_M = \gamma_R N_A \Lambda / 2\pi\mu$, leading to

$$\Lambda_{opt} = \frac{2\pi E_0}{N_A} \left(\frac{\mu \varepsilon_0 \varepsilon}{\gamma_R e} \right)^{1/2} \tag{5.12}$$

and to an optimum frequency detuning of the pump beam by

$$\delta\omega = 2\pi v_{opt} \cdot (\Lambda_{opt}^{-1}) = s I_{R0} \frac{N_D}{N_A} \frac{E_q}{E_0} . \tag{5.13}$$

It may be concluded that recording in GaAs with a moving grating has two consequences. First, under optimum conditions (v_{opt} and Λ_{opt}) the space charge field is $\pi/2$ out of phase with the interference pattern, i.e., all the space charge field is useful for promoting the energy transfer from the reference beam to the low-intensity probe beam. Secondly, the modulus of the space-charge field is increased from the value of mE_0 in the absence of fringe movement to $m(E_q/2)$ at the velocity v_{opt} (typically a five- to tenfold increase for $E_0 \approx 10 \text{ kV cm}^{-1}$).

The dependence of the gain Γ on the incident beam ratio β is interpreted by introducing the second-order terms in the expansion of the space charge field (second-order perturbation). The space-charge field takes the form [5.37]

$$E_{sc} = \tfrac{1}{2}E_{s1} \exp\left[iK(x - vt)\right] + \tfrac{1}{2}E_{s2} \exp\left[2iK(x - vt)\right] + c.c. \; .$$

5.2.2 Summary of Crystal Performance

Table 5.1 summarizes some of the properties of different photorefractive crystals that are of interest for optical signal processing applications. In addition to the data concerning the time response, recent observations have shown that, once the grating is recorded (after time τ), a differential gain equal to $\gamma_{diff} = \exp(\Gamma d/2)$ is available at very high speed (10^{-6}s and less) [5.38–40]. Application to homodyne wavefront detection is discussed in Sect. 5.3.10.

Table 5.1. Properties of some photorefractive crystals for cw incident intensity ≈ 10–100 mW cm^{-2}. τ: time response; Γ: exponential gain coefficient; R: reflectivity in 4WM. (In $KNbO_3$ [5.1] and $BaTiO_3$ [5.41], faster speeds have been demonstrated with highly reduced crystals and elevated temperature, respectively)

	$\lambda[\mu m]$	τ	$\Gamma[\text{cm}^{-1}]$	R
Ferroelectrics				
$LiNbO_3$, $KNbO_3$	0.514	seconds	10–20	1–50
$BaTiO_3$, SBN				
Nonferroelectrics				
BSO, BGO	0.568	10-100 ms	8–15	1–3
GaAs	1.06	10-100 ms	6–7	1–5

5.3 Application of Beam Coupling to Optical Computing Operations

5.3.1 Image Amplification

The large values of the gain coefficient Γ in photorefractive crystals permit the amplification of a low intensity signal beam containing spatial information (data plane) [5.42]. The optical setup for image amplification of a signal wavefront modulated by a photographic transparency is shown in Fig. 5.8. With this configuration, the energy transfer from the pump beam allows receipt of an amplified image in the detection plane. When using a photorefractive amplifier such as GaAs or BSO, an electric field is applied on the crystal and the fringe velocity is adjusted in order to receive maximum gain. However, since the spatial

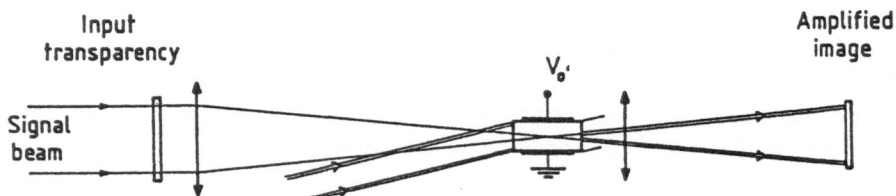

Input
transparency

Amplified
image

V_o

Signal
beam

Fig. 5.8. Optical setup for coherent image amplification. The input signal is introduced into the signal beam path as a spatially intensity modulated wavefront and amplified via 2WM

Fig. 5.9a–c. Amplified images via 2WM in photorefractive crystals. (a) ($\times 1000$) in BaTiO$_3$, $\lambda = 0.514\,\mu$m (diffusion). (From [5.43]). (b) ($\times 20$) in BSO, $\lambda = 0.568\,\mu$m (drift mode). (From [5.25]). (c) ($\times 5$) in GaAs, $\lambda = 1.06\,\mu$m (drift mode). (From [5.26])

frequency response of the photorefractive amplifiers are of the bandpass type, the difference in gain for the various spatial frequencies may be noticeable and can limit the size of the image to be amplified. Figure 5.9 shows amplified images for BaTiO$_3$ at $\lambda = 0.514\,\mu$m [5.43, 44], BSO at $\lambda = 0.568\,\mu$m [5.25] and GaAs at $\lambda = 1.06\,\mu$m [5.26]. Efficient image amplification can be performed with photorefractive crystals such as BaTiO$_3$ in which the phase-shifted volume hologram is recorded by diffusion (no applied field) with a carrier spatial frequency of the order $\Lambda^{-1} \sim 1000\,\mathrm{mm}^{-1}$. Higher values of the gain are possible in BSO and GaAs when the electric field is increased, but this would correspond to a loss in image uniformity and quality. To summarize, the main limitations of these coherent image amplifiers stem from crystal inhomogeneities and from light-induced scattering, which limit both the resolution and the minimum intensity of the image to be amplified. This light-induced scattering is due to the local fluctuations of the crystal dielectric constant, thus causing scattered waves which are efficiently amplified by 2WM with the pump beam. Clearly, in these experiments, a trade-off between the gain of the photorefractive amplifier and the signal-to-noise ratio and resolution of the amplified image has to be expected.

5.3.2 Phase Conjugation in Photorefractive Crystals

Optical phase conjugation with photorefractive crystals utilizes a four-wave mixing (4WM) interaction to reverse both the direction of propagation and the phase

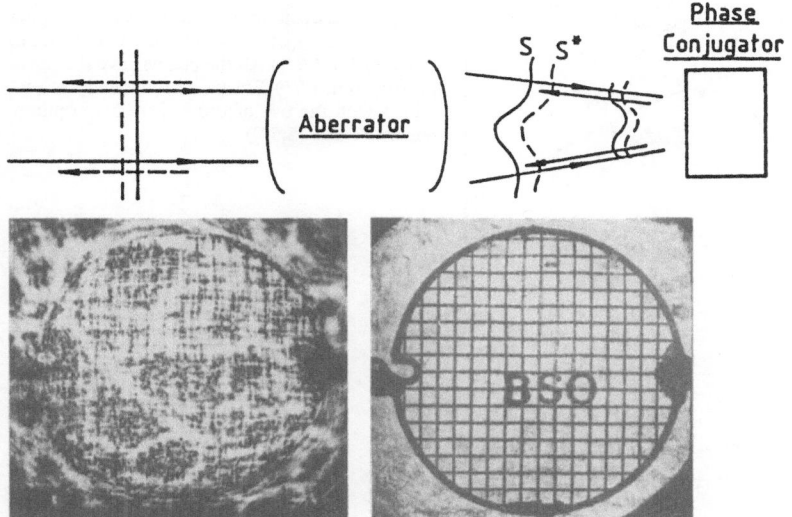

Fig. 5.10. Phase distortion compensation by wavefront reflection at a phase conjugate mirror. The distorted image (*bottom left*) is restored after travelling back through the same medium. (From [5.46])

of an arbitrary input wavefront [5.45]. Phase conjugate mirrors have many applications in problems associated with the passage of light through distorting media: the phase distortion can be removed by allowing the wavefront to travel back through this same medium (Fig. 5.10).

A second property of phase conjugate mirrors is their ability to generate a conjugate signal with an amplified intensity. Amplified phase conjugation has been observed in photorefractive crystals such as $LiNbO_3$, $LiTaO_3$, $KNbO_3$, $BaTiO_3$, and SBN [5.46–50], with typical time responses of several seconds, and in BSO [5.51] and GaAs [5.52] (10–100 ms) when recording with a moving grating. The optical configuration used for phase conjugation by nearly degenerate 4WM is presented in Fig. 5.11. In this interaction, the conditions of high reflectivity

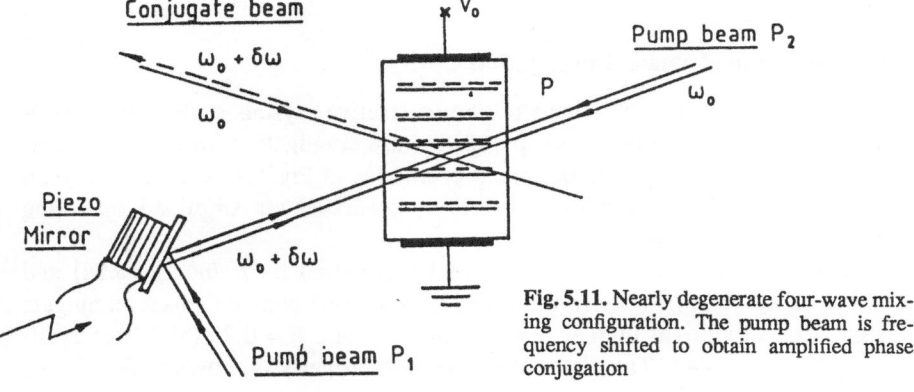

Fig. 5.11. Nearly degenerate four-wave mixing configuration. The pump beam is frequency shifted to obtain amplified phase conjugation

163

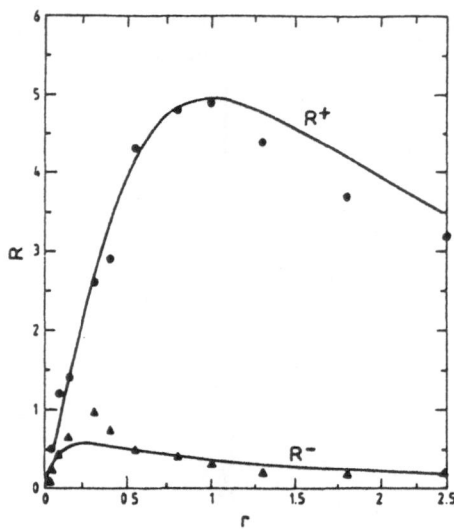

Fig. 5.12. Reflectivity of GaAs phase conjugate mirrors as a function of the pump beam ratio $r = I_{P2}/I_{P1}$. (R^-) All the beams have the same polarization. (R^+) Enhanced reflectivity is obtained with the use of crossed-polarized pump beams. (From [5.52])

solely depend on the same parameters as the exponential gain coefficient Γ previously considered in the 2WM interaction, i.e., the fringes in the crystal move at a constant velocity, and the fringe spacing is adjusted to the optimum value ($\Lambda_{opt} \approx 20\,\mu m$ for GaAs). Two important extra parameters in these 4WM interactions are (i) the pump beam ratio $r = I_{P2}/I_{P1}$ and (ii) the polarization of the "reading" beam I_{P2}. As shown in Fig. 5.12, there is a noticeable dependence of the conjugate beam reflectivity on these parameters. The lowest curve (R^-) is obtained for parallel polarizations of the interacting beams. The maximum of reflectivity ($R^- \sim 1$) obtained for the asymmetric pump beam ratio ($r < 1$) is in accordance with the coupled-mode theory developed be *Fischer* et al. [5.50]. Higher reflectivities ($R^+ = 5$) are demonstrated with crossed polarization of the pump beams (upper curve, R^+). This last configuration was first proposed by *Stepanov* et al. [5.53] as a means of taking full advantage of the two orthogonal axes of birefringence available in cubic crystals so as to provide gain for both the signal beam and the phase conjugate beam.

5.3.3 Self-Pumped Phase Conjugation

The large electro-optic coefficients of photorefractive crystals such as $BaTiO_3$ or SBN permit the realization of self-pumped phase conjugate mirrors. While classical 4WM employs two external pump beams (as in Fig. 5.12), the pump beam of a self-pumped phase conjugate mirror is generated via amplified scattering and interface reflections.

Figure 5.13a illustrates an experimental setup used by *Feinberg* [5.54] and *Rakuljic* et al. [5.55] to study the reflectivity of self-pumped phase conjugate mirrors. Reflectivities as high as $R = 0.7$ ($BaTiO_3$) and $R = 0.25$ (SBN) are available in these materials. The distortion-correcting property is shown in Fig. 5.13b.

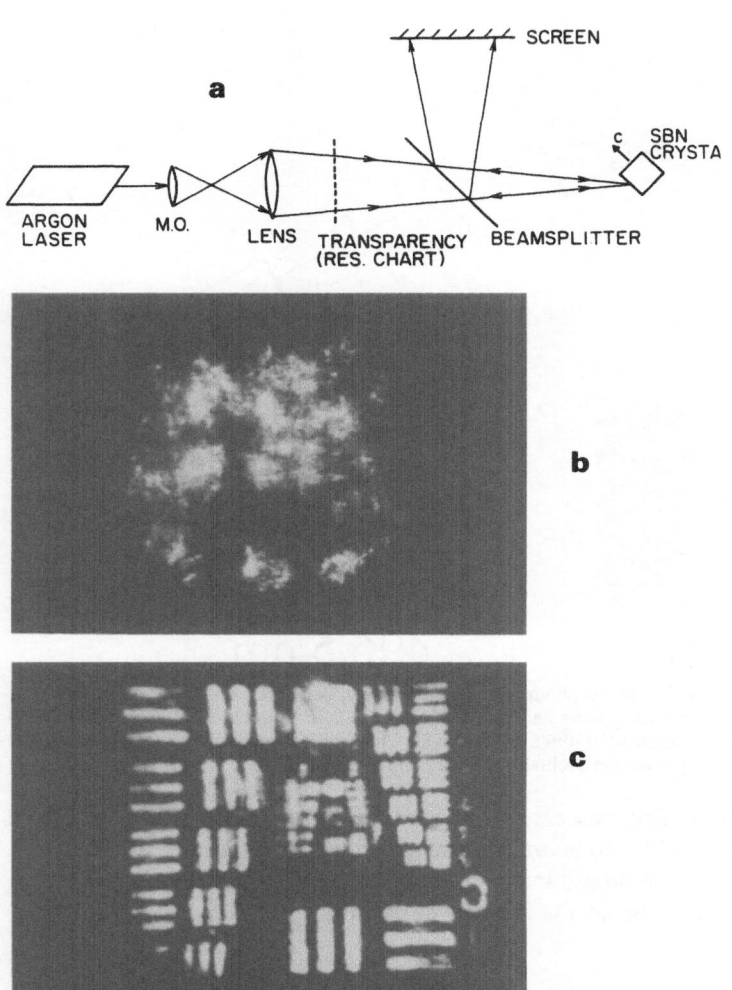

Fig. 5.13. (a) Experimental setup for studying the reflectivity of phase conjugate mirrors. The distortion-correcting property is demonstrated by inserting a phase aberrator on the signal beam path. (b) Distorted signal; (c) output image. (From [5.55])

More recently, *Weiss* et al. [5.56] and *Ewbank* [5.57] have reported the operation of double phase-conjugate mirrors. Two independent inputs to opposite sides of a photorefractive $BaTiO_3$ crystal carrying different spatial images were shown to pump the same four-wave mixing process, resulting in the phase conjugate reproduction of the two images simultaneously. The double-color-pumped photorefractive oscillator shown in Fig. 14a is an extension of the double-phase conjugate mirror: a $BaTiO_3$ crystal is pumped by two lasers of different colors [5.58]. Despite their difference in wavelength, these beams interact with each other to cause the self-generation of a common grating and two oscillation beams

165

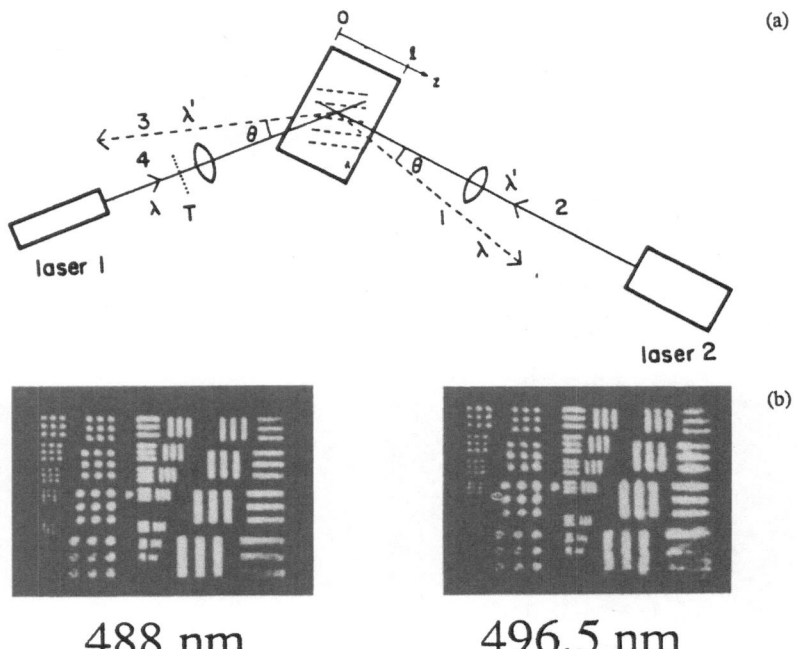

488 nm   496.5 nm

Fig. 5.14. (a)Double-color-pumped photorefractive oscillator. Two laser beams of different colors interact to generate a common grating and two oscillations of different colors. (b) Color conversion from $\lambda = 488\,\text{nm}$ (laser 1) to $\lambda' = 496.5\,\text{nm}$ (laser 2). The spatial modulation T (resolution chart) carried by beam 4 emerges on the oscillation beam 3. (From [5.58])

of different colors. This process is self-produced and permits a pictorial input (on beam 4 in Fig. 5.14a) to be transferred to beam 3 (at a different wavelength). Figure 5.14b shows the output image 3 ($\lambda' = 496.5\,\text{nm}$) when the corresponding resolution chart ($\lambda = 488\,\text{nm}$) is inserted in beam 4.

5.3.4 Laser Beam Steering/Optical Interconnections

The applications to laser beam steering and optical interconnections relies upon the use of a two-dimensionl spatial light modulator in combination with a photorefractive crystal (Fig. 5.15a). The basic principle is as follows: the pump beam interferes in the crystal with the probe beam, the direction of which is selected by the spatial light modulator (array of electro-optic shutters; for example), and after two-beam coupling, a complete energy exchange from the pump to the selected probe beam direction can be obtained by using photorefractive crystals with large gain coefficients. Therefore we can say that the pump beam has been deflected in the direction of the probe. If another direction of the probe beam is selected, the previous grating is erased, and writing of a new one deflects the pump beam in another direction. Using this principle, a new type of random-access digital laser beam deflector with large scan angles is realized.

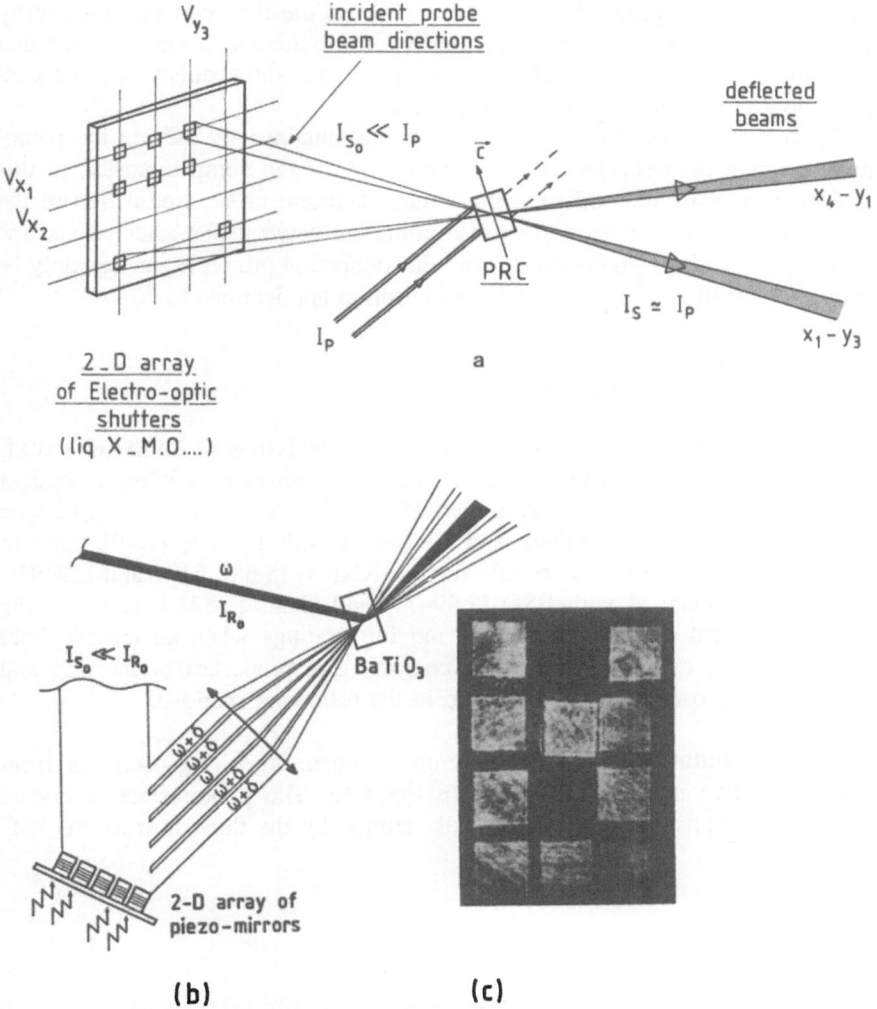

Fig. 5.15. (a) Application of the energy transfer in BaTiO₃ to 2d laser beam steering. (b) Laser beam deflection obtained by driving an array of 4 × 3 piezo-mirrors. (c) Generated pattern. The beam is randomly deflected by driving all the piezo-mirrors except the ones corresponding to the deflection positions. (From [5.59])

A practical demonstration of this principle for a limited number of positons is achieved with the experimental setup shown in Fig. 5.15b. The low intensity signal beam is expanded and reflected by an array of piezo-mirrors (4 × 3). In the focal plane of lens L, where a photorefractive BaTiO₃ crystal is placed, the pump beam and the array of signal beams interfere. Selection of one probe beam direction is achieved as follows: all of the piezo-mirrors are excited with a ramp generator except one, which corresponds to the selected direction of deflection [5.59]. Due to the Doppler shift δ induced by the moving mirrors, the interference fringes move. If $\delta \gg \tau^{-1}$ ($\tau \sim 1$ s is the time constant for energy exchange in BaTiO$_3$), the corresponding index modulation cannot be recorded due to the crystal inertia. Therefore, the probe beam, whose direction is selected by the nonexcited piezo-mirror, is amplified.

Figure 5.15c shows the experimental results obtained by driving the piezo-mirrors, where the deflected beam is about 10% of the pump intensity. In this experiment, the use of a mirror array with a temporal phase modulation of the incident wavefront enables perfect discrimination between the selected and the non-selected probe beam directions, and this deflection principle can certainly be applied for reconfigurable optical interconnection applications [5.60–64].

5.3.5 Self-Induced Optical Cavities

Due to the large gain coefficients and the large reflectivities in 2WM and 4WM, different types of self-starting oscillators can be obtained by adding an optical feedback to the photorefractive amplifiers [5.65–67]. These coherent oscillations were first observed by *Feinberg* and *Hellwarth* with BaTiO$_3$ [5.49], then in high electro-optic coefficient crystals such as KNbO$_3$ [5.68], SBN and LiNbO$_3$. They are also obtained with BSO [5.69–71] and GaAs [5.52] because of the gain enhancement due to self-induced moving gratings when an electric field is applied to the crystal. Some of the characteristic properties of the ring and phase-conjugate oscillators are reviewed in the following sections.

(a) **Ring Oscillators.** The optical setup for obtaining a ring oscillator from a photorefractive amplifier is shown in Fig. 5.16. The photorefractive crystal (GaAs) is introduced into the beam path defined by the three mirrors M$_1$ -M$_2$

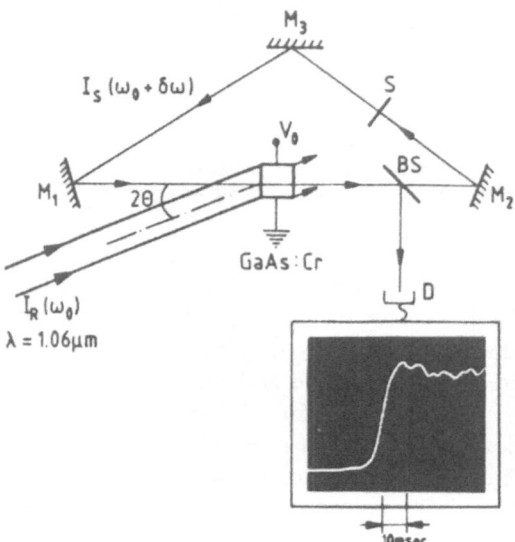

Fig. 5.16. Self-induced optical ring resonator with photorefractive GaAs

-M$_3$, and the angle between the pump beam and the M$_1$ -M$_2$ direction is chosen so as to correspond to the optimum fringe spacing for the energy transfer of the pump beam. The condition for oscillation is

$$(1 - R_{BS})R^3 \exp\left[(\Gamma - \alpha_t)d\right] \geq 1 , \tag{5.14}$$

where R and R_{BS} are the reflectivities of the cavity mirrors and beam splitter, Γ is the gain coefficient of the 2WM interaction, and α_t represents the total losses. Since the values of $\Gamma(\Gamma > 6 - 7\,\mathrm{cm}^{-1})$ considerably exceed the cavity losses ($\alpha_t \approx 3\,\mathrm{cm}^{-1}$), oscillation builds up in the cavity. The oscillation in the cavity is self-starting; the optical noise due to the pump beam is sufficient to generate a weak probe beam that is then amplified after each round trip in the cavity. The required detuning $\delta\omega$ between the pump and the cavity beam in the ring oscillator is also self-induced. In other words, from the optical noise spectrum, the crystal chooses the frequency component shifted by $\delta\omega$ that will be optimally amplified in the cavity. In photorefractive GaAs, for an applied voltage $V_0 = 5\,\mathrm{kV}$, the beam in the cavity is typically frequency-shifted 10–100 Hz for $I_{R0} = 40\,mW\,\mathrm{cm}^{-2}$ at $\lambda = 1.06\,\mu\mathrm{m}$. A specific property of these photorefractive ring oscillators is that the gain is unidirectional and only one wave is amplified in the cavity. In particular, the residual coherent retrodiffused beams due to the mirrors M$_1$, M$_2$ and M$_3$ are not amplified: after interference with the pump beam, they produce reflection-type photoinduced gratings that are not efficiently recorded in the GaAs with this configuration. The theory of oscillation in photorefractive ring resonators is developed in [5.72, 73] and, in particular, it is shown that the amount of frequency shift depends on the length of the cavity. Consequently, these photorefractive resonators may be used in a new type of interferometry, which directly converts optical pathlength changes into frequency shifts. The peculiarities of these ring cavities can also be applied for the conception of new gyroscopes based on the Sagnac effect.

The specific properties of these oscillators have also been applied to analog optical computing. In particular if an operator such as a matrix-vector multiplier is introduced into the cavity, the feedback loop permits parallel iterative algorithms to be implemented. For example, the inversion of a matrix B can be obtained by calculating the sum $\sum(I - B)^n$, each term being provided by a round trip in the photorefractive cavity [5.74].

(b) Oscillators with Phase Conjugate Mirrors. As reviewed in Sect. 5.3.2, the conjugate beam reflectivity in a 4WM interaction exceeds unity after optimization of the grating recording parameters. It is thus possible to induce an oscillation between a classical mirror and a photorefractive phase conjugate mirror. Since the first demonstration with a BaTiO$_3$ crystal [5.49], similar phase conjugate resonators have been obtained with LiNbO$_3$ and, more recently, with BSO [5.69] and GaAs crystals [5.52]. As shown in Fig. 5.17 for GaAs, the oscillation in the cavity builds up from the noise only when a frequency shift $\delta\omega$ is introduced between the pump beams. Under such conditions, the beam oscillating in the

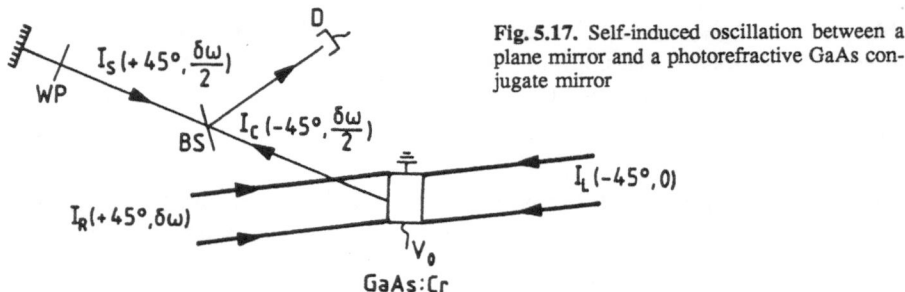

Fig. 5.17. Self-induced oscillation between a plane mirror and a photorefractive GaAs conjugate mirror

cavity is frequency shifted by $\delta\omega/2$, and this frequency shift ensures a grating moving at the optimum velocity. The oscillation is maintained even if an aberrator is placed between the mirror M and the phase conjugator crystal.

5.3.6 Optical Logic Gates and Parallel Algorithmic State Machines

In the field of digital optical computing, two-beam coupling in photorefractive materials offers unique features that suggest the implementation of single-instruction multiple data (SIMD) machines. The modules of such a machine consist of large arrays of 1-bit processors executing identical instruction streams in parallel. An important part of a (SIMD) machine is an optical logic gate array that can be addressed repeatedly, thus yielding practical realization of algorithmic state machines. *Fainman* et al. demonstrated digital optical Boolean logic such as **OR**, **AND**, **NOR** and **NOT** using two-beam coupling in photorefractive crystals [5.75]. In these materials, the nonlinear phenomena rely on the properties of the gain coefficient Γ in a classical two-beam interaction with gain. Three different effects are employed to perform optical logic operations: gain saturation, pump-beam depletion, and optically controlled two-beam coupling. An example of an OR gate is shown in Fig. 18a. The weak signal beam is amplified by the same amount when one or both pump beams A_2', A_2'' (both bearing spa-

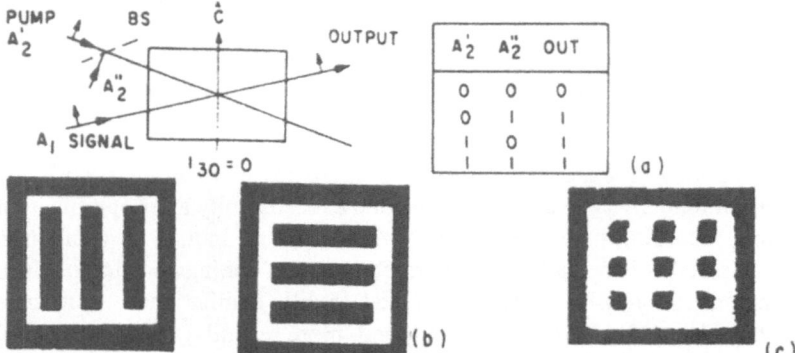

Fig. 5.18a–c. Optical logic OR by two-beam coupling in photorefractive crystals. (a) Principle of operation: high intensity level (logic 1) is obtained when either pump, A_2' or A_2'', is present. (b) Input images A_2' and A_2''. (c) Output logic $\{A_2'$ OR $A_2''\}$. (From [5.75])

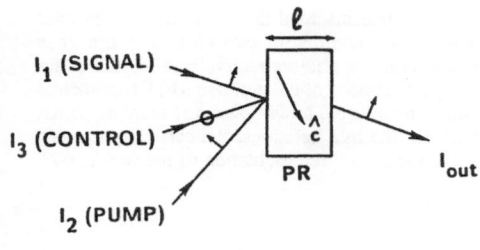

Fig. 5.19. (*Left*): Optical logic gate $\{(I_1)$ AND (I_2) AND (NOT $I_3)\}$. I_3 is a control beam that can erase the grating. (*Right*): A parallel half-adder circuit (XOR and AND) is implemented based on a combination of three BaTiO$_3$ crystals. (From [5.76])

tial information) are present with a high intensity level (logic 1). This property, related to the saturation of the photorefractive gain versus the incident pump intensity, leads to the logic operation OR. The results shown in Fig. 5.18b–c were obtained with photorefractive BaTiO$_3$ crystals and using high/low level transmittance transparencies that spatially modulate the intensities of the pump beams A_2' and A_2''.

In the interaction shown in Fig. 5.19, the signal amplification is controlled by an additional crossed polarized signal that can erase the interference grating formed by the signal and pump beams. Logic 1 (high intensity level) is obtained only when: (i) Signal and pump beams (I_1 and I_2, respectively) are in logic 1 (high intensity level) and (ii) control beam I_3 is in logic 0 (low intensity level). Consequently, this interaction provides the implementation of the logic gate $\{(I_1)$ AND (I_2) AND (NOT $I_3)\}$. A parallel AND gate $\{(A)$ AND $(B)\}$ can therefore be implemented and, combining two crystals, an exclusive OR (XOR) can also be performed as $\{(A)$ XOR $(B)\} = \{(A)$ AND (NOT $B)\}$ OR $\{($NOT $A)$ AND $(B)\}$. A detailed analysis of the dynamic range of these interactions can be found in [5.76]. The input images (two-valued intensity bit planes) were introduced in real time with the use of a microcomputer-controlled spatial light modulator. Figure 5.19 shows the experimental results for input images containing 256 bits in a 16×16 format. The real time capability of this device was used for the realization of a photorefractive algorithmic state machine (PASM) that implements a real-time sequential parallel algorithm for binary addition. The technique is to repeatedly address the parallel (AND, XOR) gates used as a half-adder circuit [5.76]. The results of the parallel addition of two sequences of five-bit numbers (i.e., two vectors) is shown in Fig. 5.20.

171

A	B	A+B
11	22	33
20	11	31
22	2	24
9	1	10
22	10	32
001011	010110	100001
010100	001011	011111
010110	000010	011000
001001	000001	001010
010110	001010	100000

(a)

Fig. 5.20a,b. An algorithmic state machine that performs a sequential algorithm for the parallel binary addition of two vectors A and B is obtained by repeatedly addressing a photorefractive parallel half-adder circuit. (a) Analog and digital versions of the problem. (b) Experimental results. *Top pictures*: Input bit planes. *Lower pictures*: Outputs of the parallel half-adder circuit for the five iterations; the carriers propagate and the last iteration gives the binary representation of the sum $A + B$. (From [5.76])

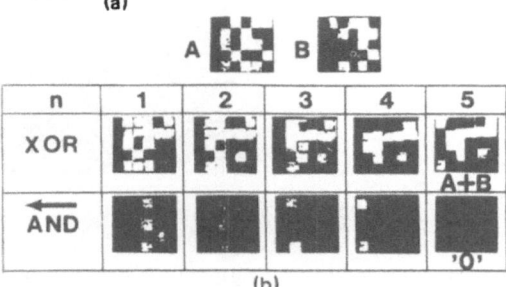

(b)

5.3.7 Image Subtraction Using a Self-Pumped Phase Conjugate Mirror Interferometer

Figure 5.21 illustrates an interferometric setup employing a self-pumped phase conjugate mirror as used by *Kwong* et al. [5.77] and *Chiou* et al. [5.78] for performing parallel image subtraction, intensity inversion and exclusive OR logic operation. The incident optical field is divided by a beam splitter BS whose amplitude reflection and transmission coefficients are r and t, respectively. For the waves propagating in the opposite directions, the amplitude reflection and transmission coefficients are r' and t'. Each of the two waves is then passed through a transparency whose intensity transmittances are T_1 and T_2. These two waves are reflected by a self-pumped photorefractive phase conjugate mirror with a nearly identical reflectivity R. The phase conjugate beams recombine interferometrically at beam splitter BS to form an output image intensity given by

$$I_{out} = I_0 R \, |t'r^*T_2 + t^*r'T_1|^2 \ .$$

From the Stokes principle of reversibility of light, it holds that

$$r't^* + t'r^* = 0 \ ,$$

and therefore

$$I_{out} = I_0 R |r'r^*|^2 \, |T_1 - T_2|^2 \ .$$

172

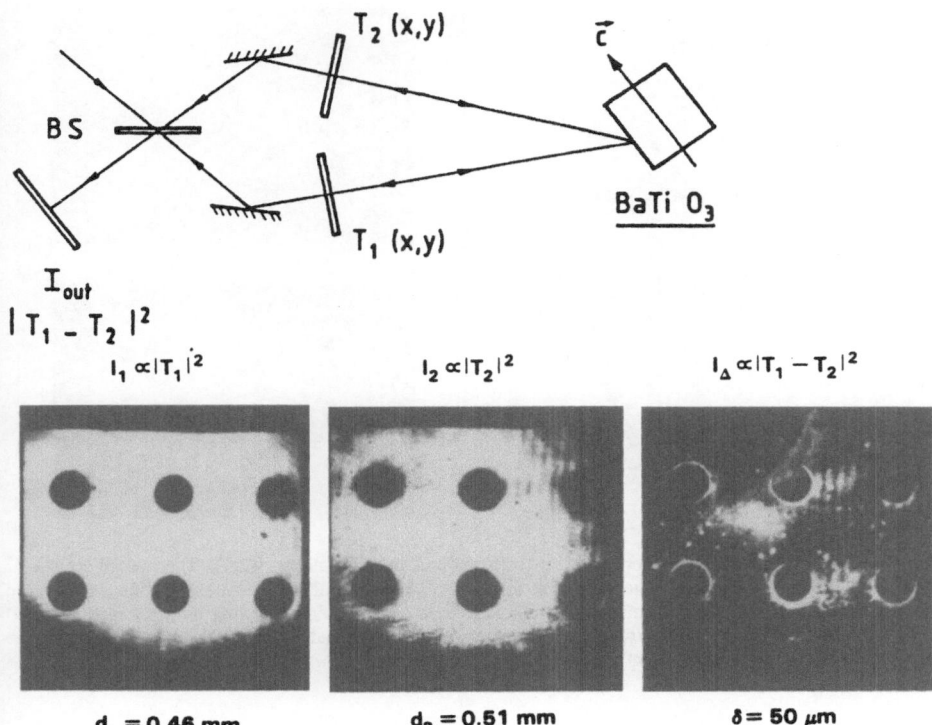

Fig. 5.21. *Top*: Real-time image intensity subtraction using a self-pumped phase conjugate BaTiO₃ crystal. *Bottom*: Experimental results of image subtraction applied to defect detection of a pair of masks. (From [5.77, 78])

Consequently, the interferometer provides an image intensity subtraction proportional to the square of the difference of the intensity transmittance functions of the two input slides. This operation represents the Boolean exclusive OR achieved in parallel between the two images T_1 and T_2. The image intensity subtraction occurs in one step. Moreover, the interferometer is only sensitive to intensity difference and is independent of the phase information of the transparencies or optical path lengths of the two arms. Experimental results of image subtraction applied to defect detection are shown in Fig. 5.21.

5.3.8 Novelty Filters

Photorefractive crystals can be employed to act as dynamic filters that detect variations in intensity or phase of a two-dimensional image or data plane. Self-pumped phase conjugation and two-beam coupling techniques were used to implement these "novelty filters". In the experiment described by *Anderson* et al. [5.79, 80], two arms of an interferometer share a common phase conjugate mirror (Fig. 5.22). A transmitting spatial light modulator imposes a phase image onto

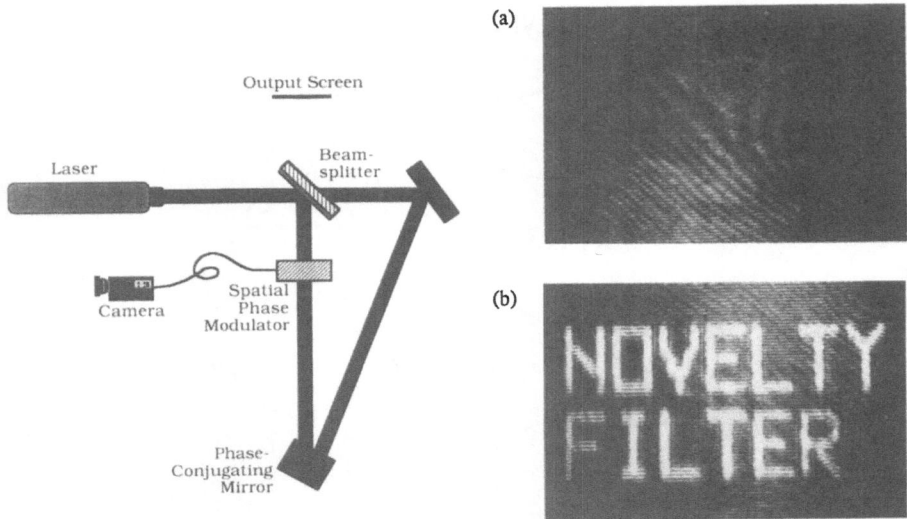

Fig. 5.22. An optical novelty filter based on self-pumped interferometry. (*Left*): The spatial light modulator imposes a phase image onto one arm of the interferometer. The output shows time-varying features of the imposed image, and returns to zero after a time exceeding the time response of the photorefractive crystal. (*Right*): Experimental results: (a) steady state; (b) output taken just after the image is fed into the light valve. (From [5.80])

the optical beam in one of the interferometer arms. The phase conjugate mirror guarantees that, in steady state, the output of the beam splitter is zero, independent of the relative lengths of the arms. However, when the phase information in one arm suddenly changes, the recombined fields at the beam splitter no longer interfere destructively, thus yielding nonzero intensity at the output. The output returns to its original state after a delay governed by the response time of the phase conjugate mirror.

More recently, *Cronin-Golomb* et al. [5.81] and *Ford* et al. [5.82] have demonstrated novelty filters using signal depletion due to noise amplification (fan-out) in a crystal of $BaTiO_3$. Depletion on an input signal is achieved by amplified scattering from crystal defects (Fig. 5.23a). As the input signal varies, the fanning hologram written in the crystal adjusts to the change according to its photorefractive response time, and the changing image components become brighter as deamplification is disturbed. Figure 5.23b shows edge detection of a model car obtained by moving the object (the input signal) in front of the crystal; only the edges are changed in intensity, resulting in a transient edge-enhanced output image.

5.3.9 Image Convolution and Correlation

Dynamic cross-correlation or spatial convolution with a classical Fourier transform lens configuration can be achieved by two- or four-wave mixing of optical

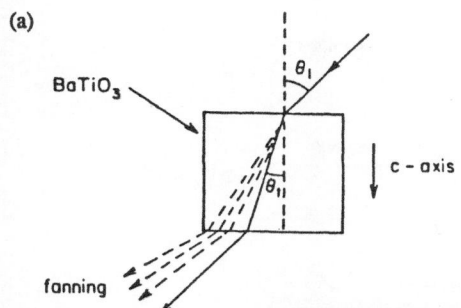

(a)

BaTiO$_3$

θ_1

c - axis

θ_{11}

fanning

Fig. 5.23. (a) Optical novelty filter using amplified scattering in BaTiO$_3$. A time-varying input signal results in a transient deamplified output. (b) Output signal showing edge enhancement of a model car. The car was moved forward slightly after the signal was completely depleted so that only the edges were changed in intensity. (From [5.81, 82])

(b)

fields in photorefractive crystals. Figure 5.24 illustrates a four-wave mixing geometry as used by *White* and *Yariv* [5.83]. All the beams have the same wavelengths, and the amplitudes $u_1(x, z)$, $u_2(x, z)$ and $u_4(x, z)$ in the outer focal planes are Fourier transformed by propagating to the common focal plane. The transform fields mixed in the photorefractive crystal are as follows:

$$U_1 = \mathrm{FT}\{u_1(x, z)\} \; ; U_2 = \mathrm{FT}\{u_2(x, z)\} \; ; U_4 = \mathrm{FT}\{u_4(x, z)\} \; .$$

The backward wave generated by 4WM in the crystal, $u_3(x, z)$ evaluated at a distance f from lens L, is

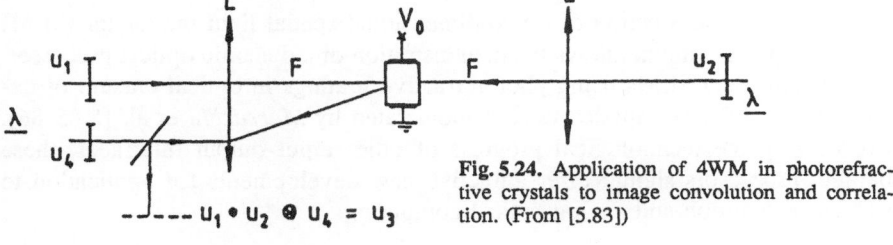

L

V_0

L

u_1

F

F

u_2

λ

u_4

λ

$u_1 \bullet u_2 \circledast u_4 = u_3$

Fig. 5.24. Application of 4WM in photorefractive crystals to image convolution and correlation. (From [5.83])

175

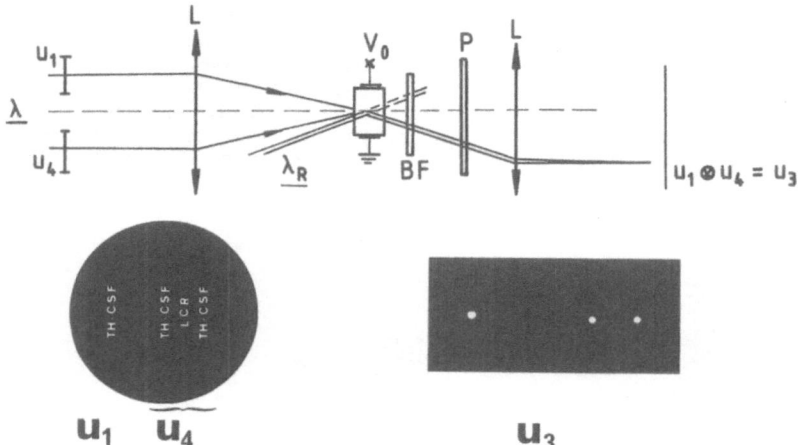

Fig. 5.25. Application of 2WM to image convolution or correlation. BF: blocking filter for λ_R; P: polarizer for noise filtering. (From [5.84])

$$u_3(x_0, z_0) = \alpha_0 u_1(-x, -z) \cdot u_2(-x, -z) \times u_4(-x, -z) ,$$

where α_0 depends on the amplitude of the photoinduced index modulation, and $\cdot$ and $\times$ denote respectively the product and the convolution product of the optical field. Figure 5.25 ilustrates another configuration based on 2WM in BSO [5.84]. The interference pattern recorded in the Fourier plane of lens L is read out by an auxiliary low power laser (He-Ne; $\lambda_R = 633$ nm). The thickness of the crystal implies that this readout beam of different wavelength has to be positioned at the correct Bragg angle to obtain the maximum intensity in the diffracted cross-correlation peak.

Experimental results using BSO as a dynamic matched filter are shown in Fig. 5.25 (typical response time $\tau = 50$ ms for 1 mW incident intensity on the photorefractive crystal). The Bragg angular selectivity of the phase volume holo-gram may limit the number of pixels that can be processed, and further stud-ies should quantitatively estimate the capabilities of these architectures for high capacity parallel optical processing of analog or digital images. In these exper-iments, an auxiliary beam incident on the photorefractive crystal may also be used to modify the output of the processor in real time; this provides a means of weighting the correlation product in favor of specified spatial frequencies. Moreover, the introduction of a two-dimensional spatial light modulator (SLM) for real-time data input, allows the demonstration of a dynamic optical processor. Two-dimensional SLMs using photorefractive gratings in optical erasure or en-hanced self-diffracted modes were demonstrated by *Marrakchi* et al. [5.85, 86]. Considering the technological progress of other input–output interfaces, these parallel processors should offer attractive new developments for application to pattern recognition and analog optical computing.

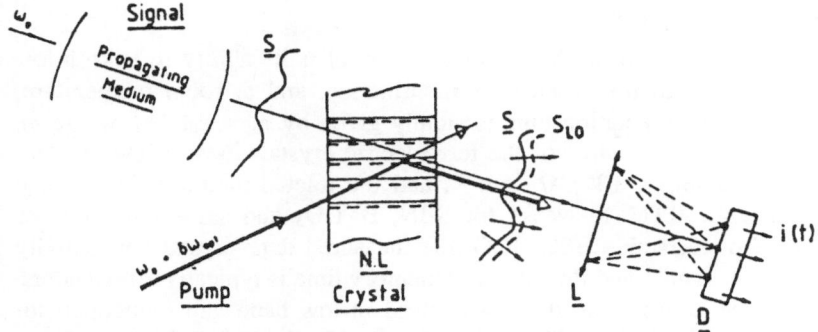

Fig. 5.26. Two-wave mixing in a nonlinear photorefractive crystal applied to coherent homodyne detection of a complex wavefront. The incident arbitrary wavefront carrying spatial and temporal information (S) is perfectly phase-matched with the local oscillator wave S_{LO} that originates from the pump diffracted at the dynamic grating. (From [5.39])

5.3.10 Coherent Homodyne Detection

The two-wave mixing interaction with gain can be used to perform homodyne detection of arbitrary optical wavefronts as proposed by *Hamel de Montchenault* and *Huignard* [5.39]. Consider a complex wavefront which carries temporal and spatial information (i.e. a wavefront diffracted by a target object, a spatial light modulator, etc.). This wave interferes with the pump beam to create a complex dynamic hologram in the volume of the photorefractive crystal. The self-diffraction of the pump beam generates a wave perfectly in phase at every point with the incident signal wavefront (Fig. 5.26). Therefore, the self-diffracted wave can be considered as a local oscillator wave in a homodyne detection scheme involving an arbitrary signal wavefront. This local oscillator wave has the properties required for this purpose: (i) Its phase is perfectly matched to the incident wavefront; (ii) its amplitude can be much larger than that of the incident signal wave provided that a large gain coefficient is achieved in the interaction; (iii) its optical frequency is equal to that of the incident signal wave. When this prevails, the minimum detectable signal corresponds to the quantum limit.

5.4 Optical Memories with Photorefractive Crystals

Dynamic photorefractive crystals potentially offer two key features for application to parallel information storage. First, high storage capacity (diffraction limit $\lambda = 10^{12}$ bits/cm^3) is available. Multiple data planes can be recorded and selectively retrieved using the Bragg selectivity of volume holograms. Secondly, information retrieval by association can be implemented when nonlinear photorefractive-based optical feedback is provided to the memory.

5.4.1 Multiple Image Storage

One of the peculiarities of a photorefractive crystal is its ability to store information in the dark. In the absence of illumination, and at room temperature, the crystal dielectric relaxation time is simply given by $\tau_{di} = \varepsilon_0 \varepsilon / \sigma_0$, where σ_0 is the dark conductivity. Most of the ferroelectric crystals have a low conductivity in the dark, $\sigma_0 < 10^{-18} \Omega^{-1} cm^{-1}$, and the related memory times range from 10 hours for KTN, to weeks for SBN, $BaTiO_3$ and several months for $LiNbO_3$. In photorefractive BSO-BGO, the measured dark crystal conductivity is about $10^{-14} \Omega^{-1} cm^{-1}$ and the observed memory time is typically 10–20 hours. The dark storage time is considerably reduced in low band gap semiconductor materials sensitive in the near IR and is typically 10^{-14} s in InP:Fe.

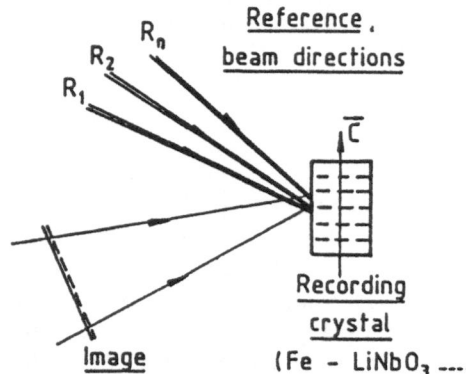

Fig. 5.27. Multiple image storage in the volume of a photorefractive crystal by angular coding of the reference beam direction

The superposition of many holograms in the same crystal volume is accomplished, for example, by varying the angle of incidence of the reference beam (Fig. 5.27), each individual hologram being associated with a well-defined angle. Selective reconstruction of the superimposed holograms is achieved because efficient diffraction occurs only when the hologram is addressed at the right Bragg angle. This multiplexing technique requires a photorefractive crystal with an asymmetric erasure cycle (for example $LiNbO_3 - 0.1\%$ Fe^{3+}) in order to prevent erasure of previously recorded data pages during the recording of a new one in the same crystal volume. At present, up to 500 holograms have been superimposed and fixed in a cube of $LiNbO_3$ crystal, giving a total capacity of 0.5 Gbits cm^{-3} [5.87]. A method for selective erasure of any information block, based on a coherent image subtraction technique was demonstrated by *Huignard* et al. [5.88].

The possibility of electrically controlled volume hologram writing in $LiNbO_3$ crystals has been demonstrated by *Petrov* et al. [5.60]. Due to the large electro-optic effect in this crystal, it is possible to control the Bragg conditions for image reconstruction by the bias voltage applied to the crystal. The independent reconstruction of two or three holograms written in the same crystal under different voltages is demonstrated in [5.60]. The maximum efficiency is reached at the same voltage as used for the recording.

5.4.2 Associative Memories

Associative memory systems that use holographic data bases and phase conjugate mirrors to provide generative optical feedback, thresholding and gain have been recently reported by *Soffer* et al. [5.89, 90], *Anderson* et al. [5.91] and *Wagner* et al. [5.92]. The principles of information retrieval by association using parallel optical techniques, and in particular those based on holographic principles, were recognized early on by various authors [5.93, 94]. However, these first approaches were limited in their ability to faithfully reconstruct the output object from a partial input because of the large crosstalk which results when multiple objects are holographically stored in the memory. Nonlinear elements such as photorefractive crystals can now overcome these problems since they introduce optical feedback and gain, thus improving the selectivity and the stability of the memory. The principle of a holographic associative memory is shown in Fig. 5.28. Only a single hologram is used in this configuration, and it is simultaneously addressed by the object as well as by the conjugate reference beam, the latter beam acting as the key that unlocks the associated information. A photorefractive $BaTiO_3$

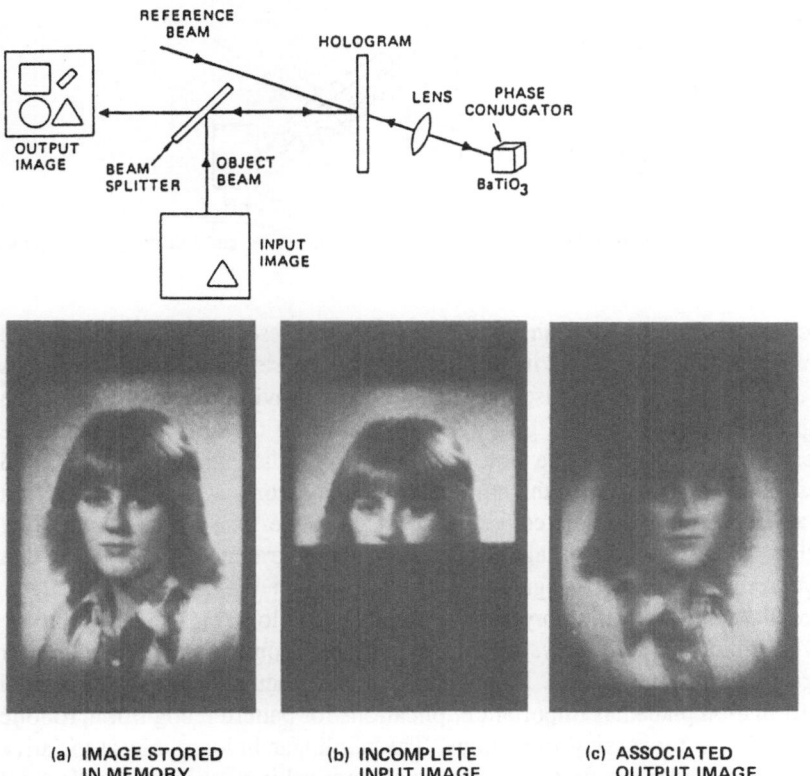

(a) IMAGE STORED IN MEMORY   (b) INCOMPLETE INPUT IMAGE   (c) ASSOCIATED OUTPUT IMAGE

Fig. 5.28. Associative holographic memory. Complete object image reconstruction from a partial input image. (From [5.89, 90])

phase conjugator is used both for reference beam retroreflection and for gain and thresholding. This provides the necessary nonlinearity, emphasizing only the strongly correlated signals. The demonstration of total image reconstruction of an object image when only a partial image addresses the system is also shown in Fig. 5.28. The illumination of the hologram by part of the object generates a diffracted beam propagating in the original direction of the reference beam. This beam is then phase-conjugated and amplified by four-wave mixing in a photorefractive $BaTiO_3$ crystal. When this readout beam impinges on the hologram, it is diffracted and recreates the initial object beam. This recreated beam contains all the information originally recorded in the hologram memory. This principle can be extended to different objects superimposed in the hologram memory by angular coding of the reference beam directions [5.89, 95].

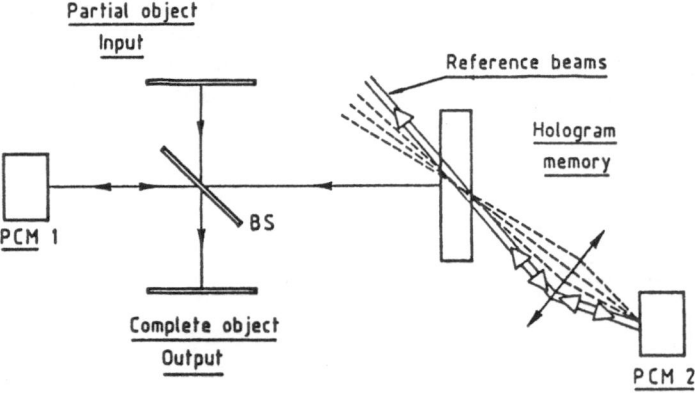

Fig. 5.29. Optical associative memory using a phase conjugate resonator and information storage in a volume holographic memory. (From [5.90])

A schematic implementation of such a nonlinear associative memory based on a hologram placed in a cavity formed by two phase conjugating mirrors is illustrated in Fig. 5.29. The phase conjugate mirrors provide beam retroreflection with gain and thresholding and give a self-alignment of the object and reference wavefronts with respect to the hologram memory. The optical feedback and the thresholding effects due to the nonlinear mirrors favor the strongly correlated signals and force the system to converge to a stable state. The steady-state output signal thus consists of the image stored in the holographic memory and that which presents the highest degree of correlation with the input image. Real-time modification of the memory is also possible if holograms are stored in the volume of a photorefractive crystal. This would be a requirement for adaptive or learning behavior of the system. This ability to reconstruct an image from a partial input information plane has important implications for pattern recognition, robotic vision and image processing operations. The nonlinear holographic associative memory also constitutes a first step towards the optical implementation of neural networks. This model is based on the feasibility of distributed and interconnected memory elements with a nonlinear feedback, and is thus analogous, in many aspects, to optical holography.

5.5 Concluding Remarks

We have presented a description of the photorefractive effect and its potential applications to some operations for use in the field of optical processing. At present, the basic physical mechanisms leading to the photoinduced index in different crystals are well identified, but, despite the great interest in applications, little is known about the charge transport processes or the species responsible for the photorefraction in these crystals. Therefore, continued research into the microscopic origins of the photorefractive effect is essential for further optimization of the materials through impurity doping and thermal treatments. It is hoped that crystals with much higher sensitivity and displaying larger photoinduced nonlinearities in the visible and near -IR wavelength range will be developed in the future. Several of the existing materials already work at millisecond speed with low power visible and near-IR lasers. Moreover, we have shown that the existence of a spatial phase shift between the incident fringe pattern and the photoinduced index modulation leads to the amplification of low intensity wavefronts by means of the energy transfer from a pump beam in two-wave or four-wave mixing interactions. The amplification factor depends on several recording parameters such as the grating period, the beam ratio of the interfering waves and the applied electric field, as well as on the intrinsic material parameters. To summarize, photorefractive crystals certainly have most of the characteristics needed for an initial demonstration of parallel optical processing devices. These capabilities will motivate further studies on materials, algorithms and architectures for a variety of new applications using optical interactions in all-optical and hybrid systems.

References

5.1   P. Günter, J.P. Huignard (eds.): *Photorefractive Materials and Their Applications I*, Topics Appl. Phys., Vol. 61 (Springer, Berlin, Heidelberg 1988)
5.2   P. Günter: Phys. Rep. **93**, 200 (1982)
5.3   H. Rajbenbach, J.P. Huignard: Proc. of the Summer School on Optical Computing, Edinburgh, August 1988
5.4   N.V. Kukhtarev, V.M. Markov, S.G. Odulov, M.S. Soskin, V.L. Vinetskii: Ferroelectrics **22**, 949 (1979)
5.5   L. Young, W.K.Y. Wong, M.L. Thewalt, W.D. Cornish: Appl. Phys. Lett. **24**, 264 (1974)
5.6   F. Micheron: Ferroelectrics **18**, 153 (1978)
5.7   G.C. Valley, M.B. Klein: Opt. Eng. **22**, 704 (1983)
5.8   H. Kogelnik: Bell Syst. Tech. **48**, 2909 (1969)
5.9   J. Feinberg, D. Heinman, A.R. Tanguay, R.W. Hellwarth: J. Appl. Phys. **51**, 1297 (1980)
5.10   R.R. Neurgaonkar, W.K. Cory, J.R. Oliver, M.D. Ewbank, W.F. Hall: Opt. Eng. **26**, 292 (1987)
5.11   P. Günter, F. Micheron: Ferroelectrics **18**, 27 (1978)
5.12   R.A. Mullen, R.W. Hellwarth: J. Appl. Phys. **58**, 40 (1985)
5.13   F.P. Stronhkendl, R.W. Hellwarth: J. Appl. Phys. **62**, 2450 (1987)

5.14   A.M. Glass, A.M. Johnson, D.H. Olson, W. Simpson, A.A. Ballman: Appl. Phys. Lett. **44**, 948 (1984)
5.15   M.B. Klein: Opt. Lett. **9**, 350 (1984)
5.16   J.P. Herriau, D. Rojas, J.P. Huignard, J.M. Bassat, J.C. Launay: Ferroelectrics **75**, 271 (1987)
5.17   G.C. Valley: IEEE J. QE–**19**, 1637 (1983)
5.18   G. Le Saux, A. Brun: IEEE J. QE–**23**, 1680 (1987)
5.19   C.T. Chen, D.M. Kim, D. Von Der Linde: IEEE J. QE–**16**, 126 (1980)
5.20   L.K. Lam, T.Y. Chang, J. Feinberg, R.W. Hellwarth: Opt. Lett. **6**, 475 (1981)
5.21   A.L. Smirl, G.C. Valley, R.A. Mullen, K. Bohnert, C.D. Mire, T.F. Boggess: Opt. Lett. **12**, 501 (1987)
5.22   J.M.C. Jonathan, G. Roosen, Ph. Roussignol: Opt. Lett. **13**, 224 (1988)
5.23   A.L. Smirl, G.C. Valley, K.M. Bohnert, T.F. Boggess, Jr.: IEEE J. QE–**24**, 289 (1988)
5.24   J.C. Fabre, J.M.C. Jonathan, G. Roosen: J. Opt. Soc. Am. B. **5**, 1730 (1988)
5.25   H. Rajbenbach, J.P. Huignard, B. Loiseaux: Opt. Commun. **48**, 247 (1983)
5.26   B. Imbert, H. Rajbenbach, S. Mallick, J.P. Herriau, J.P. Huignard: Opt. Lett. **13**, 327 (1988)
5.27   V. Kondilenko, V. Markov, S. Odulov, M. Soskin: Opt. Acta **26**, 238 (1979)
5.28   A. Marrakchi, J.P. Huignard, P. Günter: Appl. Phys. **24**, 131 (1981)
5.29   J.P. Huignard, A. Marrakchi: Opt. Commun. **38**, 249 (1981)
5.30   R.W. Hellwarth: J. Opt. Soc. Am. **67**, 1 (1977)
5.31   J.P. Huignard, J.P. Herriau: Appl. Opt. **24**, 4285 (1985)
5.32   S.I. Stepanov, M.P. Petrov: Opt. Commun. **53**, 292 (1985)
5.33   X. Gan, S. Ye, Y. Sun: Opt. Commun. **66**, 155 (1988)
5.34   J. Kumar, G. Albanese, W.H. Steier: J. Opt. Soc. Am. B **4**, 1079 (1987)
5.35   R.B. Bylsma, A.M. Glass, D.H. Olson: Electron. Lett. **24**, 360 (1988)
5.36   G.C. Valley: J. Opt. Soc. Am. B. **1**, 868 (1984)
5.37   Ph. Refregier, L. Solymar, H. Rajbenbach, J.P. Huignard: J. Appl. Phys. **58**, 45 (1985)
5.38   G. Hamel de Montchenault, B. Loiseaux, J.P. Huignard: Appl. Phys. Lett. **50**, 1794 (1988)
5.39   G. Hamel de Monchenault, J.P. Huignard: J. Appl. Phys. **63**, 624 (1988)
5.40   R.B. Bylsma, A.M. Glass, D.H. Olson: Electron. Lett. **24**, 362 (1988)
5.41   D. Rytz, M. Klein, R.A. Mullen, R.N. Schwartz, G.C. Valley, B.A. Wechsler: Appl. Phys. Lett. **52**, 1759 (1988)
5.42   V. Markov, S. Odulov, M. Soskin: Opt. Laser Technol. **11**, 95 (1979)
5.43   F. Laeri, T. Tschudi, J. Albers: Opt. Commun. **47**, 387 (1983)
5.44   Y. Fainman, E. Klancnik, S.H. Lee: Opt. Eng. **25**, 228 (1986)
5.45   J. Feinberg: *Optical Phase Configuration*, ed. by R.A. Fisher (Academic, London 1983) pp. 417–443
5.46   J.P. Huignard, J.P. Herriau, Ph. Aubourg, E. Spitz: Opt. Lett. **4**, 21 (1979)
5.47   S. Odulov, M. Soskin, V. Vasuetsov: Opt. Commun. **32**, 183 (1980)
5.48   P. Günter: Opt. Lett. **7**, 10 (1982)
5.49   J. Feinberg, R.W. Hellwarth: Opt. Lett. **5**, 519 (1980)
5.50   B. Fischer, M. Cronin–Golomb, J.O. White, A. Yariv, R. Neurgaonkar: Appl. Phys. Lett. **40**, 863 (1982)
5.51   H. Rajbenbach, J.P. Huignard, Ph. Refregier: Opt. Lett. **9**, 558 (1984)
5.52   H. Rajbenbach, J.P. Huignard: Opt. Lett. **14**, 78 (1989)
5.53   S.I. Stepanov, M.P. Petrov: Opt. Acta **31**, 1335 (1984)
5.54   J. Feinberg: Opt. Lett. **7**, 486 (1982)
5.55   G.A. Rakuljic, K. Sayano, A. Yariv, R.R. Neurgaonkar: Appl. Phys. Lett. **50**, 10 (1987)
5.56   S. Weiss, S. Sternklar, B. Fischer: Opt. Lett. **12**, 114 (1987)
5.57   M. Ewbank: Opt. Lett. **13**, 47 (1988)
5.58   S. Sternklar, B. Fischer: Opt. Lett. **12**, 711 (1987)
5.59   D. Rak, I. Ledoux, J.P. Huignard: Opt. Commun. **49**, 302 (1984)
5.60   M.P. Petrov, S.I. Stepanov, A.A. Kamshilin: Opt. Commun. **21**, 297 (1977)
5.61   G. Pauliat, J.P. Herriau, A. Delboulbé, G. Roosen, J.P. Huignard: Opt. Soc. Am. B **3**, 306 (1986)
5.62   J.W. Goodman, F.J. Leonberger, S.Y. Kung, R.A. Athale: Proc. IEEE **72**, 850 (1984)
5.63   P.D. Henshaw: Appl. Opt. **21**, 2323 (1984)
5.64   P. Yeh, A.E.T. Chiou, J. Hong: Appl. Opt. **27**, 2093 (1988)
5.65   M. Cronin–Golomb, B. Fischer, J.O. White, A. Yariv: IEEE J. QE- **20**, 12 (1984)
5.66   P. Yeh: Opt. Soc. Am. B **2**, 1924 (1985)

5.67   D.Z. Anderson, R. Saxena: J. Opt. Soc. Am. B **4**, 164 (1987)
5.68   G. Pauliat, P. Günter: Opt. Commun. **66**, 329 (1988)
5.69   H. Rajbenbach, J.P. Huignard: Opt. Lett. **10**, 137 (1985)
5.70   J.P. Huignard, H. Rajbenbach, Ph. Refregier, L. Solymar: Opt. Eng. **24**, 586 (1985)
5.71   P. Pellat–Finet, J.L. de Bogrenet de la Tocnaye: Opt. Commun. **55**, 305 (1985)
5.72   A. Yariv, S.K. Kwong: Opt. Lett. **10**, 454 (1985)
5.73   M.D. Ewbank, P. Yeh: Opt. Lett. **10**, 496 (1985)
5.74   H. Rajbenbach, Y. Fainman, S.H. Lee: Appl. Opt. **26**, 1024 (1987)
5.75   Y. Fainman, C. Guest, S.H. Lee: Appl. Opt. **25**, 1598 (1986)
5.76   H. Rajbenbach: J. Appl. Phys. **62**, 4675 (1987)
5.77   S.K. Kwong, G.A. Rakuljic, A. Yariv: Appl. Phys. Lett. **48**, 201 (1986)
5.78   A.E. Chiou, P. Yeh, M. Khoshnevisan: Nonlinear Optical Image Processing for Potential Industrial Applications. Opt. Eng. **27**, 385 (1988)
5.79   D.Z. Anderson, D.M. Lininger, J. Feinberg: Opt. Lett. **12**, 123 (1987)
5.80   D.Z. Anderson, M.C. Erie: Opt. Eng. **26**, 435 (1987)
5.81   M. Cronin–Golomb, A.M. Biernachi, C. Lin, H. Kong: Opt. Lett. **12**, 1029˙ (1987)
5.82   J.E. Ford, Y. Fainman, S.H. Lee: Opt. Lett. **13**, 856 (1988)
5.83   J. White, A. Yariv: Appl. Phys. Lett. **37**, 5 (1980)
5.84   L. Pichon, J.P. Huignard: Opt. Commun. **36**, 277 (1981)
5.85   A. Marrakchi, A.R. Tanguay, Jr., J. Yu, D. Psaltis: Opt. Eng. **24**, 124 (1985)
5.86   A. Marrakchi: Opt. Lett. **13**, 655 (1988)
5.87   D.L. Staebler, W. Burke, W. Philips, J.J. Amodei: Appl. Phys. Lett. **26**, 182 (1975)
5.88   J.P. Huignard, J.P. Herriau, F. Micheron: Appl. Phys. Lett. **26**, 256 (1975)
5.89   B.H. Soffer, G.J. Dunning, Y. Owechko, E. Marom: Opt. Lett. **11**, 118 (1986)
5.90   Y. Owechko, G.J. Dunning, E. Marom, B.H. Soffer: Appl. Opt. **26**, 100 (1987)
5.91   D.Z. Anderson: Opt. Lett. **11**, 56 (1986)
5.92   K. Wagner, D. Psaltis: Appl. Opt. **26**, 5061 (1987)
5.93   R.J. Collier, K.S. Pennington: Appl. Phys. Lett. **8**, 44 (1966)
5.94   D. Gabor: IBM J. Res. Dev. **13**, 156 (1969)
5.95   A. Yariv, S. Kwong, K. Kyuma: Appl. Phys. Lett. **48**, 114 (1986)

6. All-Optical Guided-Wave Devices for Switching and Routing

Y. Silberberg and P.W. Smith

With 14 Figures

There is a growing interest in the technology of optical switching and routing for applications in future communication networks. One reason for this interest is the major commitment that is being made to fiber optics for data transmission. The demand for more and more sophisticated services continues to push the development of communications networks with an increasingly high bit rate capacity. We are now at the stage where lightwave communication networks are limited not by the fiber capacity, but by the electronics used for switching, routing and signal processing (Fig. 6.1). Because all-optical switching devices can function in a time range inaccessible to electronics, it is tempting to suggest that ultrafast all-optical signal-processing elements will be used in future communications networks to alleviate the electronics "bottleneck" [6.1]. In this chapter, we will review the status of one important class of all-optical devices – all-optical guided wave devices – and indicate areas where future progress can be expected.

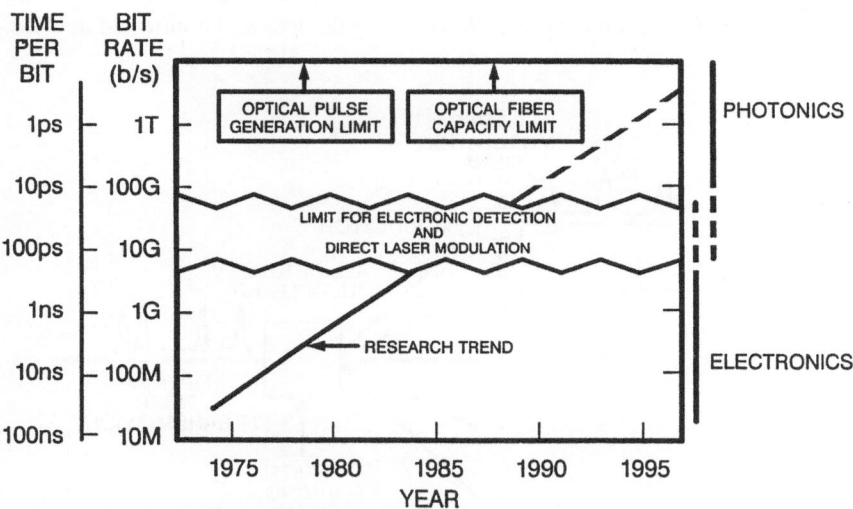

Fig. 6.1. Progress in lightwave communications systems. Experimental systems are now at the point where bit rates are high enough to place severe demands on the electronic parts of the system. As the trend towards still higher bit rates continues, it will become more and more attractive to perform some of the signal processing using optics in place of electronics

Springer Series in Electronics and Photonics, Vol. 30
Nonlinear Photonics Editors: H.M. Gibbs · G. Khitrova · N. Peyghambarian
© Springer-Verlag Berlin Heidelberg 1990

6.1 Introduction

6.1.1 Speed or Parallelism?

Speed and parallelism are often cited as potential advantages of optical signal processing. However, specific systems are likely to take advantage of one or the other of these features. Highly parallel systems must operate with low switching power per optical element so that the total power required to operate an array of them will not exceed attainable system values and will not cause severe thermal problems. Because of fundamental trade-offs in nonlinear materials and devices, this implies that switching devices designed for low-power parallel array operation will tend to have slow response times: fast devices will require high switching power. Most of the chapters in this book deal with parallel systems and devices. In this chapter, we will look at the other side of the coin: What do we obtain if we give up parallelism for speed?

Because of the relatively high power required for a fast all-optical switching operation, these devices are likely to be used in applications where only a relatively small number of switching operations are required. One example is shown in Fig. 6.2. Here, a high bit-rate optical communications system utilizes all-optical time division multiplexing to increase the capacity of the system.

In many cases, the all-optical processing required is of a simple nature. For example, a thresholding operation is called for in some novel communication schemes [6.2, 3]. An all-optical thresholding device would eliminate an electronic operation that could limit the capacity of the system. More ambitious are devices in which an optical signal can be actually routed into different channels. Routing can be determined by either the signal intensity itself or by a second controlling signal. We can define, accordingly, self-switching devices and controlled devices.

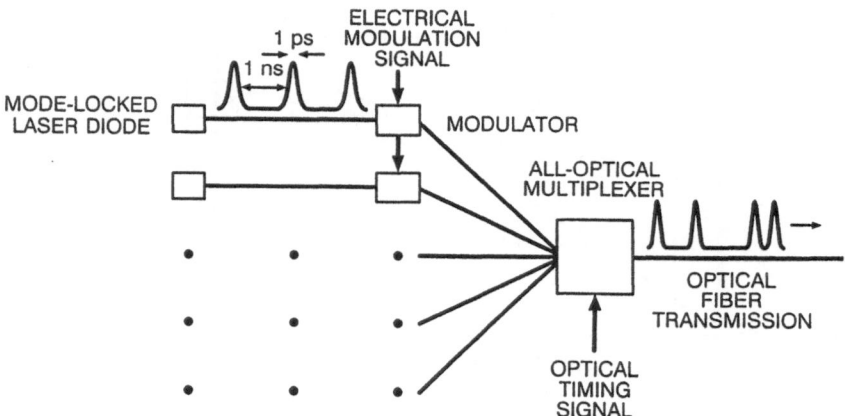

Fig. 6.2. Proposed all-optical time division multiplexer. Although the modulation of an individual pulse train is at a low enough rate to be performed electro-optically, the multiplexing element must switch on a picosecond time scale, and thus an all-optical element must be used

Controlled devices can be coherent, i.e. phase sensitive, or incoherent, where the signal and control can have different wavelengths.

6.1.2 Material Requirements

All the devices we will discuss require a material with an intensity-dependent refractive index. We assume an ideal Kerr law dependence:

$$n = n_0 + n_2 I \ . \tag{6.1}$$

The index of refraction n cannot, of course, increase indefinitely with the intensity I. Some saturation effects are expected. However, in transparent materials such as silica and glass, saturation is seldom observed.

Most of the all-optical switching devices that have been studied to date have fallen into two categories. The bistable Fabry-Perot devices can be made small in size but require materials with large optical nonlinearities. Semiconductor materials are probably the most attractive candidates, but they have response times that are limited by carrier recombination times. They also suffer from thermal effects due to large absorbed power in a small volume and a limited wavelength range of operation. Waveguide devices typically have interaction lengths of a centimeter or more. They can thus be made with materials having smaller nonlinearities and suffer less from problems of thermal effects and limited wavelength range. Because they are traveling wave devices, they can respond in times much shorter than their transit time, if materials with sufficiently fast nonlinearity are used.

Although fast turn-on times have been demonstrated with many devices, rapid recovery is usually impeded by thermal and carrier recombination effects. For many applications, thermal heating turns out to be a major problem. A useful figure of merit, F, can be defined as follows [6.4]:

$$F = \frac{\text{fast index change required to produce switching}}{\text{resultant thermal index change}} \ .$$

With this definition, the figure of merit for a number of nonlinear optical materials is given in Table 6.1.

It can be seen that optical glass, although it possesses a relatively small non-linearity, is so transparent that it has the highest figure of merit of any nonlinear

Table 6.1. Figure of merit for nonlinear optical materials

Material	$\lambda[\mu m]$	$\tau[ps]$	F
GaAs	1.06	1	$< 3 \times 10^3$
GaAs MQW	0.85	100	3×10^2
PTS	1.9	1	3×10^4
CdSe-doped glass	0.53	30	7×10^4
SF59 glass	1.06	1	5×10^5
Silica fiber	1.3	1	4×10^7

optical material. That fact, coupled with its ultrafast nonlinear response and its capability to be made into optical fibers, has made glass an attractive material for all-optical devices. The high optical power required for switching, however, is a serious impediment to practical applications.

The material issue is perhaps the most important factor in successful implementation of all-optical devices. Detailed discussion of this issue is beyond the scope of this chapter. However, it is clear that no ideal material has emerged so far, and extensive research into this problem is still required.

6.1.3 Fiber and Planar Waveguide Devices

We will concentrate in this chapter on nonlinear devices in planar waveguides and fibers. The confinement of light in the small core area, and its diffractionless propagation over long distances, increase the efficiency of the nonlinear interaction and allow the use of relatively weak nonlinearities. Many of the proposed waveguide devices have been first realized in a fiber-optic form. While still in a less-developed state compared with optical fibers, planar waveguide structures offer several possibilities that are difficult to realize in fibers [6.5]. Planar structures are more suitable for implementation in different nonlinear materials systems such as semiconductors, organics and certain glasses. Even materials which are difficult to process into waveguide structures can be considered, for example, as cladding layers. Most attractive, however, is the possibility of generating planar waveguide structures that have unique nonlinear characteristics, structures that would have been impossible in a fiber device.

Waveguide geometry is also attractive because it matches nicely to the linear part of the communication system where the signal is transmitted in a waveguide. It would be desirable to affect the signal – switch it or route it – while it was propagating in a waveguide structure. To benefit from the waveguide geometry without being limited by transit times, it is desirable to have devices that operate in a traveling-wave, or "pipeline", mode. This implies that while the time required for the signal to pass through the device may be relatively long, several consecutive operations can be performed as fast as the nonlinearity and dispersion effects allow; thus the rate can be very high.

Since the magnitude of the nonlinear index change is usually small, it can be treated as a perturbation over the index profile that defines the guided wave structure. The main effect of this perturbation is to change the optical phase of the propagating mode. Most proposed waveguide devices make use of this nonlinear phase shift. A mode with a transverse field distribution $\varepsilon(x, y)$ which propagates a length L in the waveguide will acquire a nonlinear phase shift [6.5]

$$\delta\phi = k_0 n_2 L P / A_{\text{eff}} , \tag{6.2}$$

where $k_0 = 2\pi/\lambda$, P is the total power in the mode, and the effective mode area is

$$A_{\text{eff}} = \frac{\left\{\int [\varepsilon(x,y)]^2 da\right\}^2}{\int [\varepsilon(x,y)]^4 da},$$

(6.3)

where the integration is across the transverse dimensions of the waveguide.

In this chapter, we will describe several types of all-optical traveling-wave waveguide devices: some of them have been demonstrated and some are still proposals. We can distinguish between several classes of devices. Interferometric devices, which are a nonlinear analog of electro-optic devices, are reviewed in Sect. 6.2. Some traveling wave devices exhibit unique nonlinear characteristics which are useful for switching; two of these are discussed in Sect. 6.3. Finally, in Sect. 6.4, we discuss some ideas that do not fall into the previous categories. We will not consider devices with feedback, such as bistable devices, which are not compatible with pipeline geometry. We want to stress that even without feedback, traveling wave devices offer a rich spectrum of nonlinear behavior, although bistability is not possible.

6.2 Interferometric Devices

6.2.1 Mach-Zehnder Interferometers

Perhaps the simplest nonlinear devices from a conceptual point of view are the several types of Mach-Zehnder devices which have been proposed [6.6–8]. These all-optical devices operate essentially like their electro-optic counterparts, with the difference that an optically induced phase shift rather than an electro-optic one affects their output. Some of these devices are shown schematically in Fig. 6.3. Simple nonlinear Mach-Zehnder interferometers are two-port self-

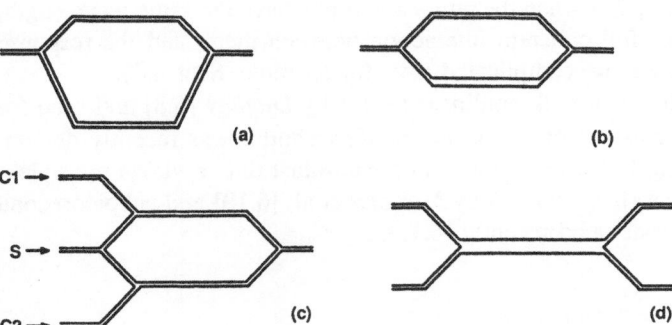

Fig. 6.3a–d. Some interferometric switching devices. Self-switching devices based on a Mach-Zehnder structure are shown in (a) and (b). A nonlinear phase shift results from an uneven length (a) or different mode sizes (b) of the two arms. The interferometer shown in (c) can be used for controlled switching, with signal in S and control pulses introduced into the arms $C1$ and $C2$. The nonlinear switch in (d) is based on an intensity-dependent phase shift between two guided modes in a dual-mode guide

switching devices, in which asymmetry is due to a difference in optical path-length or different mode size [6.6] leads to intensity-induced phase shifts between the two arms. These devices are characterized by a sinusoidal transfer function

$$P_{out} = P_{in} \sin^2(\phi_0 + \delta\phi_1 - \delta\phi_2) \,, \tag{6.4}$$

where ϕ_0 is the low power phase difference between the two arms and $\delta\phi_i$ are the nonlinear phase shifts in each of the two arms. Clearly, when the arms are identical there is no self-switching effect.

Adding two input channels to a symmetric Mach-Zehnder results in a very versatile logic gate [6.7]. In this case, separate control signals should be separable from the probe signal by using either different wavelengths or different polarizations. The proposed device shown in Fig. 6.3 also requires an electro-optically induced phase bias.

6.2.2 Kerr Shutters

The all-optical Kerr shutter is based on optically induced birefringence that rotates the polarization of a signal wave. A linearly polarized pump field with wavelength λ_p induces birefringence in an otherwise isotropic media. In particular, when the nonlinearity originates from an electronic process, a signal wave at λ_s experiences an index change for its two polarization directions related by

$$\delta n_\perp = \frac{2}{3}\delta n_\parallel = \frac{2k_0 n_2 PL}{(3A_{eff})} \,. \tag{6.5}$$

This results in a relative phase shift which is 1/3 the magnitude of the nonlinear phase shift for linearly polarized light. The transmission of the weak signal wave through a polarizer depends sinusoidally on the birefringence induced by the strong pump field. The signal wave is separated from the pump by a suitable filter. The polarization rotation can be viewed as an interference between orthogonal waves. Note, however, that when the pump and probe have the same wavelength, one must include the full coherent interaction between them, and the response becomes significantly more complicated (see, for example, Sect. 4.3).

First demonstrated as a bulk nonlinear device by *Duguay* [6.8] and used for ultrafast sampling measurements, the optical Kerr shutter was recently demonstrated in a fiber form [6.9]. Its application for ultrafast time division multiplexing/demultiplexing has been studied by *Morioka* et al. [6.10] and subpicosecond gating has been demonstrated recently [6.11].

6.2.3 Dual-Mode Interferometers

A third interferometric device that has been proposed [6.12] and recently demonstrated [6.13, 14] for power-dependent routing is shown in Fig. 6.3d. This is the all-optical equivalent of an electro-optical device known as the BOA switch [6.15]. In this device, the central region supports two guided modes, which are excited with equal amplitudes through each of the input arms. The device routes the signal according to the nonlinear phase shift induced between the two guided modes of the central region. The two modes have different shapes, and they experience different nonlinear phase shifts. In a self-switching scheme, each mode carries the same power P. The nonlinear phase shift between them is given by [6.12]

$$\delta\phi = k_0 n_2 L P \int (\varepsilon_0^4 - \varepsilon_1^4) da , \tag{6.6}$$

where the symmetric and antisymmetric mode profiles ε_0 and ε_1 are assumed to be normalized. The response of this device for relatively low powers is, again, a sinusoidal function of the input power. Equation (6.6) holds only when the phase shift $\delta\phi$ is smaller than the ordinary phase difference $(\beta_0 - \beta_1)L$, β_i being the propagation constant of mode i. A very different behavior is expected at the high power regime, as described in the next section. The experiment in [6.13] uses dual mode elliptic core fiber to demonstrate the effect. A control "pump" at 1.06 μm induced switching of 0.633 μm signal.

6.3 Dual-Mode Switches

6.3.1 Nonlinear Directional Couplers

While the interferometric devices discussed in the previous section are basically all-optical analogs of electro-optical devices, there is a second family of devices which are based on a unique nonlinear process which can be used to obtain several interesting and potentially useful characteristics. The nonlinear directional coupler (NLDC) is the member of this family that has attracted the most attention [6.16–40]. Other seemingly different nonlinear processes can be shown to yield the same responses. These include polarization switching in birefringent fibers [6.41], two-mode coupling in dual-mode fibers [6.12], and polarization switching in periodic fiber filters [6.42]. The basic underlying process in all these systems is the nonlinear interaction between two guided modes. Here, we will not derive the theory behind these devices. The interested reader is referred to the original paper by *Jensen* [6.16], who first proposed the NLDC, and to various other publications that study other aspects of these devices [6.17–40]. We will discuss only the main results.

In the most straightforward application as a switch, a coupler of one coupling length is used. Light is coupled into one of the two waveguides, denoted here as waveguide 1. As the light propagates down the coupler, low intensity signals are coupled from the input guide to the other waveguide, while high intensity signals bias the coupler and remain in the input guide, as shown schematically in Fig. 6.4. The exact analysis results in the following expression for the light intensity in waveguide 1 as a function of the length z and power P:

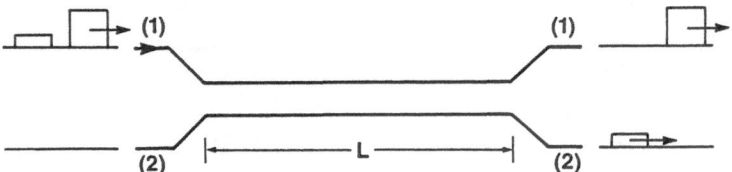

Fig. 6.4. Nonlinear directional coupler as an all-optical switch. In a one coupling-length device, low input intensity signals are coupled to guide 2, while high intensity signals are switched to guide 1

$$\frac{P_1}{P} = \frac{1}{2}[1 + cn(\pi z/2L_c|P^2/P_c^2)] \ . \tag{6.7}$$

Here, cn $(\phi|m)$ is the Jacobi elliptic function, L_c is the coupling length and P_c is a critical power, which is given by

$$P_c = \lambda A_{\text{eff}}/n_2 L_c \ . \tag{6.8}$$

The critical power is the power needed for a 2π nonlinear phase shift in a single waveguide of the same dimensions as the coupler waveguides and length L_c. Note that the critical power is inversely proportional to the coupling length.

The nonlinear coupler and an electro-optically controlled coupler behave differently. In the electro-optic devices, the index is changed by an electric field of a given spatial distribution. The complex behavior of the nonlinear coupler results from the interdependence of the light distribution and the index change. At low light intensities, light which is fed into one waveguide couples back and forth between the two waveguides as it propagates through the device with a periodicity of $2L_c$. At somewhat higher power levels, the light still oscillates between the guides; complete power transfer is still obtained, but the period of the oscillations grows with the power. As the power approaches the critical value P_c, the period approaches infinity; at P_c the signal is evenly divided between the two waveguides. At higher values of power, the propagation is periodic again; the period decreases now as power is increased, and the power transfer between the guides is no longer complete. Figure 6.5 shows the response of a NLDC of length L_c to power. A relatively abrupt transition is obtained at powers of the order of P_c. The usefulness of such a response for switching is evident.

When a long device of several coupling lengths is used, a very sharp transition is obtained at powers around P_c. The longer the coupler, the sharper this transition, as shown in Fig. 6.6. This effect can be used to generate a high gain amplifier [6.18, 36]. This sensitive region can also be applied to form a phase-controlled switch, in which a weak control signal can route an intense input [6.19].

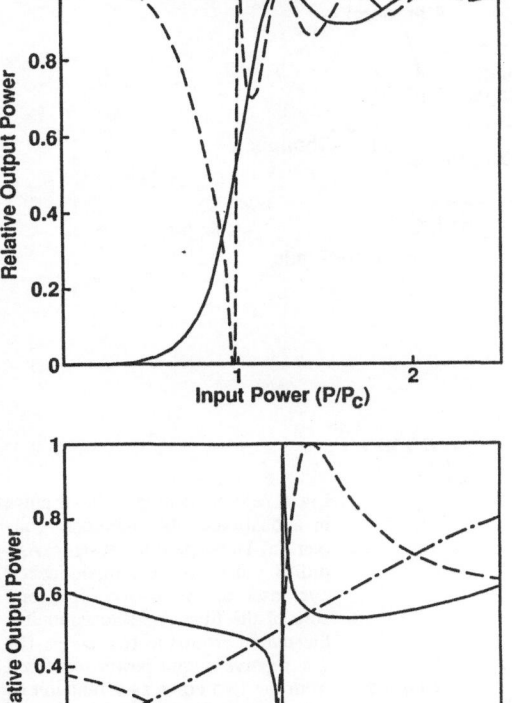

Fig. 6.5. The calculated relative output from guide 1 (the input guide) of nonlinear directional couplers of length L_c (*solid line*) and $2L_c$ (*dashed line*) as a function of the input power. The power is shown in units of the critical power value P_c

Fig. 6.6. The sharp transition in transmission near the power value $P = P_c$ for couplers which are 2 (*dash-dotted line*), 3 (*dashed line*), and 4 (*solid line*) coupling lengths long. Note the expanded power scale

6.3.2 Experimental Nonlinear Directional Coupler Studies

Experimental demonstrations of switching in NLDCs have been made in dual-core fibers [6.31–35] and in semiconductor devices [6.36–40]. The fiber experiments require much higher power than semiconductor versions; however, they are free from the absorption and thermal nonlinearities that complicate the interpretation of the semiconductor experiments.

Dual-core fiber couplers can be designed to give a wide range of coupling lengths. Although longer coupling length reduces the switching power, it is impractical to consider very long couplers because of the very tight tolerances that will be imposed on the core symmetry. Long couplers will also be extremely sensitive to external perturbations and particularly to bends along the fiber.

Extensive studies of dual-core fiber couplers have been carried out in the USSR by *Maier* and co-workers [6.31–33]. They have demonstrated the switch-

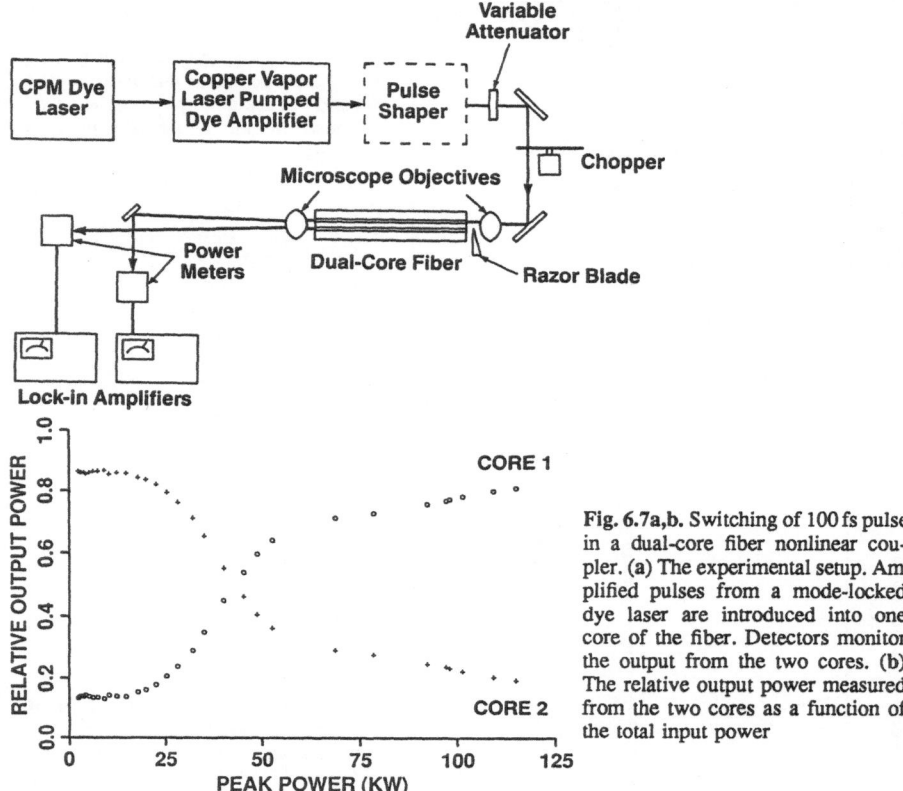

Fig. 6.7a,b. Switching of 100 fs pulses in a dual-core fiber nonlinear coupler. (a) The experimental setup. Amplified pulses from a mode-locked dye laser are introduced into one core of the fiber. Detectors monitor the output from the two cores. (b) The relative output power measured from the two cores as a function of the total input power

ing of picosecond light pulses in couplers that were typically 1 m long. More recently, a Bellcore group has demonstrated switching of 100 fs pulses at a wavelength of 620 nm in a 5 mm fiber coupler [6.35]. This short coupler was necessary because dispersion would have affected the short pulse in a longer coupler. Figure 6.7a shows the experimental arrangement, and Fig. 6.7b depicts the measured results. It is obvious from Fig. 6.7b that the switching obtained was not complete. This is a general problem with ultrafast switching which is discussed in Sect. 6.4.1. Note also the high peak powers, of the order of 50 kW, that were needed, although it is important to remember that the switching energies are still reasonably low because of the short duration of the pulses. This experiment is the fastest switching experiment to date, and it illustrates the problem and potential of all-optical switching.

6.3.3 Polarization Switching and Other Analogs

As mentioned earlier, a number of other dual-mode systems have been shown to be complete analogs of the NLDC. Table 6.2 is a convenient "translation" table that shows the analogies between the various systems.

Table 6.2. Comparison between different implementations of dual-mode switches

Type	Length	Input	Output (Low power)	Output (High power)
Nonlinear Directional Coupler	1 coupling length	Guide 1	Guide 2	Guide 1
Nonlinear Birefringent Fiber	1/2 beat length	cw pol.	cw pol.	cw pol.
Nonlinear Periodic Coupler	1 coupling length	Mode 1	Mode 2	Mode 1

Two of the proposed devices use polarization rather than spatial modes to obtain switching. In these versions, the outcoming signal is switched between polarization states. These can then be used to switch or route the signal using a proper polarization component. In one experiment, *Trillo* et al. [6.41] used a slightly birefringent fiber to demonstrate switching between two circular polarization states. A 50 cm device showed evidence of switching at powers on the order of 400 W. More recently, the same group used a periodic fiber polarization filter to demonstrate switching between two linear polarizations [6.42]. The latter device has the advantage that it can be made longer (and hence switch at lower powers) with less sensitivity to perturbation compared with a coupler or a birefringent fiber. The price for the lower power requirement is a limitation on the operating wavelength.

6.3.4 Nonlinear Waveguide Junctions

So far, we have discussed two-mode devices that are uniform, or at most periodic, along the propagation direction z. Here, we consider structures which vary along z. In particular, we look at waveguide junctions, in which two single-mode waveguides converge to form a single two-mode waveguide. Such a structure is known to behave in two distinct ways, depending on its symmetry [6.43]. It was recently used to generate a new digital electro-optic switch [6.44]. *Silberberg* and *Sfez* [6.45] have found that asymmetries introduced by nonlinear index changes can drive a symmetric junction to behave like an asymmetric one. Particularly useful is a geometry where the light is coupled to the two-mode region and is propagating into diverging guides. This system has unique routing capabilities: a high intensity signal at the two-mode region will tend to route itself to one of the two diverging waveguides with good contrast ratio. This is because the symmetric mode of propagation is extremely unstable, and small perturbations are sufficient to break the symmetry and to direct the entire signal to one branch. This behavior is demonstrated in Fig. 6.8: an intense signal is coupled to the fundamental mode of the two-mode section. Without any perturbation, this signal splits evenly between the two output guides, as shown in the center simulation. Slight asymmetry in the field pattern can route the signal to the left or to the right. An addition of a small quantity of the higher order mode is sufficient to

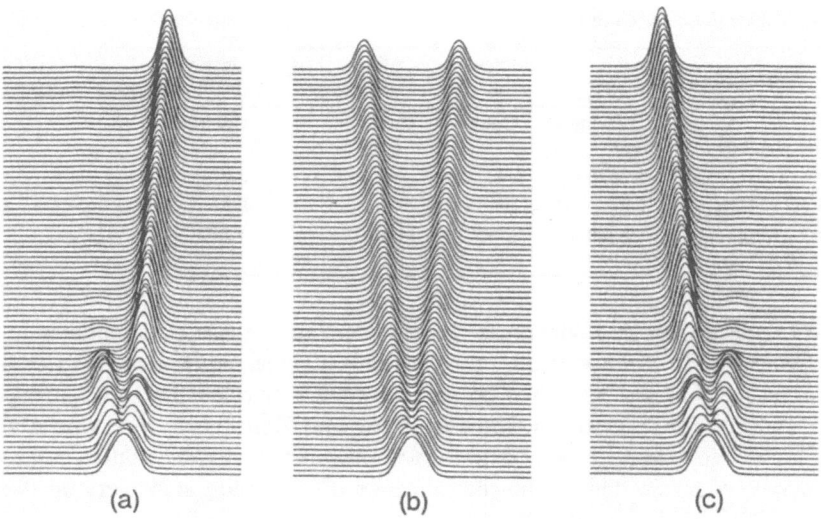

(a)   (b)   (c)

Fig. 6.8a–c. Simulations of controlled switching in a nonlinear symmetric Y-junction. The input in (b) is the pure fundamental mode of the base of the junction. In (a) and (c) there is a small higher order (antisymmetric) component, 1/200 in intensity, with a relative phase of 0 and π, respectively. The intensity of the light is such that the total nonlinear phase shift along the junction is of order 10π

induce switching. The almost complete switching shown in Fig. 6.8 is obtained with a control signal which is less than 1% of the switched signal. Figure 6.9 demonstrates the robustness of this process: the switch is insensitive to the precise value of relative phase between the two modes, and the state is stable also against perturbations of the intensity. This should be contrasted with the high sensitivity of the controlled NLDC device which has been proposed [6.19, 36]. Nonlinear waveguide junctions can provide the basis for phase or intensity controlled switching devices. As with most nonlinear devices, the power requirements can be reduced by making the device longer. In the nonlinear junction, this means reducing the angle between the guides. Generally speaking, a nonlinear phase accumulation of a few times π over the interaction length is required in order to reach this binary regime.

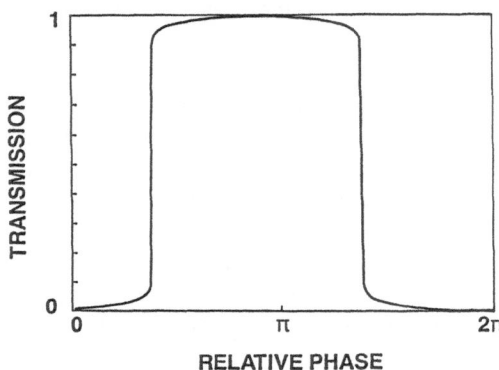

Fig. 6.9. The output through the left arm of the Y-junction shown in Fig. 6.8 as a function of the relative phase between the symmetric signal and antisymmetric control field

6.3.5 The Pulse Breakup Problem

Nonlinear optical devices respond to the instantaneous intensity of the light. Ultrashort pulses have a varying intensity: they are usually bell-shaped with a time dependence that can be approximated by a Gaussian or $sech^2$ function. When an ultrashort pulse is switched in a nonlinear device such as a NLDC, its weaker wings and its peak are going to be switched in different ways. The result is a reduction in the contrast ratio of the switch when averaged over the entire pulses. For example, we show in Fig. 6.10 the response of an ideal NLDC switch for cw light together with the calculated average response to a $sech^2$ pulse. This phenomenon explains the incomplete switching observed in Fig. 6.7b. A more direct demonstration of this point is shown in Fig. 6.11, which shows the autocorrelation traces of the light emerging from the two cores of the NLDC in the experiment depicted in Fig. 6.7 [6.35]. The peak power was about $2P_c$. The traces verify that the center of the pulse is switched to waveguide 1, while the wings are emerging from waveguide 2.

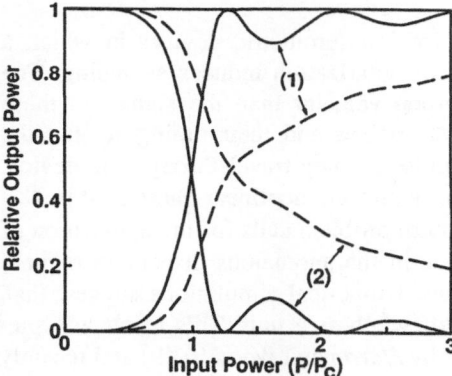

Fig. 6.10. The response of a nonlinear directional coupler switch to cw light (*solid line*) and to a $sech^2$ shaped pulse (*dashed line*)

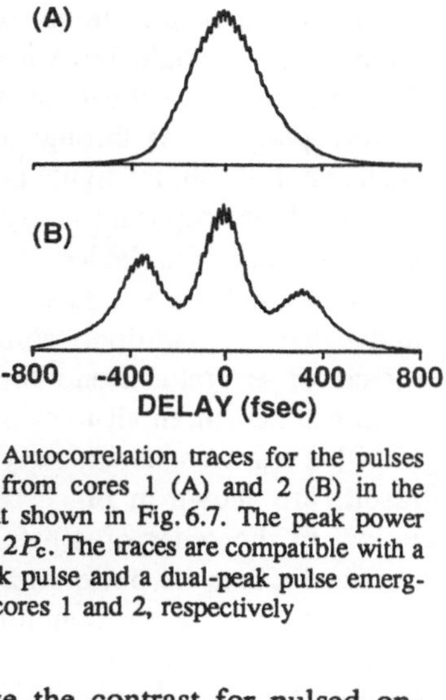

Fig. 6.11. Autocorrelation traces for the pulses emerging from cores 1 (A) and 2 (B) in the experiment shown in Fig. 6.7. The peak power was about $2P_c$. The traces are compatible with a single-peak pulse and a dual-peak pulse emerging from cores 1 and 2, respectively

Several proposals have been made to improve the contrast for pulsed operation. The most straightforward solution is to use a square optical pulse. A perfect square pulse should reconstruct the ideal cw response. Pulse shaping in the picosecond and femtosecond regime has been demonstrated, including the generation of square pulses [6.46]. In a recent experiment [6.47] a 540 fs long square pulse, with rise and fall times of 100 fs, was used with a dual-core fiber coupler to generate the switching characteristics shown in Fig. 6.12. Note the improved extinction ratio and the lower switching intensity of the square pulse compared with a regular bell-shaped pulse.

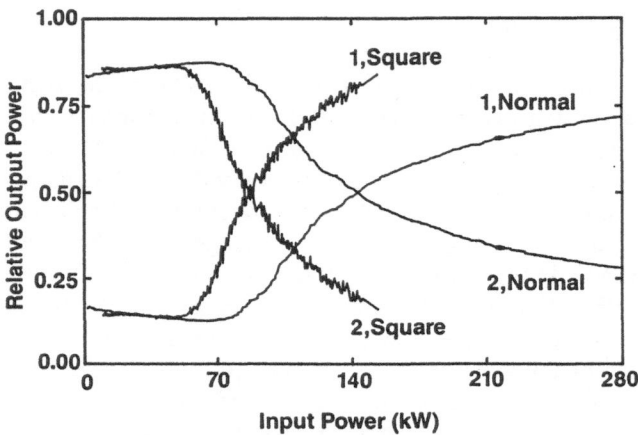

Fig. 6.12. Improved switching in a nonlinear coupler with a square optical pulse. The experimental setup is shown in Fig. 6.7a, with the pulse shaper in place. Plots show the relative output power from the two cores (marked 1 and 2) for 100 fs normal bell-shaped input pulses (Normal) and for 540 fs square pulses (Square)

A second solution is suitable only for interferometric devices in which a control pulse of a different wavelength or polarization induces switching. The control pulse would have a different group velocity than the signal. In these devices, one can adjust the length of the pulses and their timing so that the control pulse slides through the signal pulse as they travel through the device. As a result, the entire signal pulse acquires a uniform nonlinear phase shift [6.48].

The third proposal to overcome the pulse problem calls for the application of solitons. Here, the devices should operate in the anomalous dispersion regime. Solitons are launched as the input pulses. Numerical simulations suggest that, under certain conditions, solitons tend to switch as a unit. This result was predicted for several soliton interferometers by *Doran* and *Wood* [6.49] and recently demonstrated in an all-fiber Sagnac interferometer [6.50, 51]. The experiment in [6.50] is shown schematically in Fig. 6.13a, and the theoretical and experimental results are shown in Fig. 6.13b. Solitons were generated at 1.5 μm, with a duration on the order of 400 fs. Switching of 93% of the total pulse power was obtained when the pulse energy was increased to 46 pJ. Solitons were also predicted to improve the switching characteristics of nonlinear couplers by *Trillo* et al. [6.52]. The improvement occurs when the soliton period is on the order of the coupling length.

6.4 Other Nonlinear Switching Phenomena

6.4.1 Nonlinear Parametric Processes and All-Optical Switching

The use of nonlinear gain processes can significantly lower the power requirements for all-optical switching. We will discuss two recent demonstrations of such effects that use Brillouin gain [6.53] and modulational instabilities [6.54].

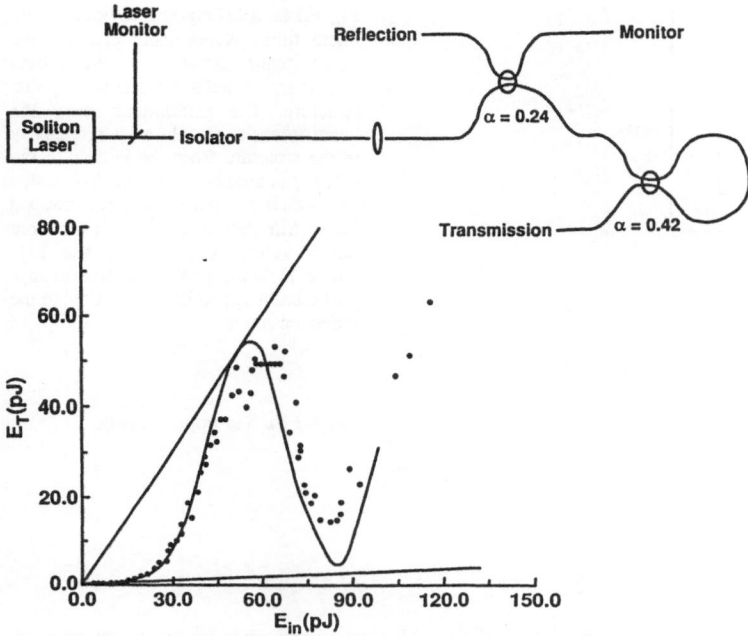

Fig. 6.13a,b. Soliton switching in a nonlinear Sagnac interferometer (after [6.50]). (a) Experimental setup. The all-fiber interferometer is the loop connected through a 42% directional coupler. At low powers, most of the power is reflected back into the input fiber. The reflection and the transmission from the "loop mirror" are monitored at the marked ports. (b) The measured (*dots*) and calculated (*solid line*) energy transmitted by the loop mirror as a function of the input pulse energy. Up to 93% of the input energy can be switched. The two straight lines are the limits of low-power and full transmission

A narrow-band pump with frequency ω_p can amplify a counterpropagating signal wave at a lower frequency ω_s if the frequency difference $\Omega = \omega_p - \omega_s$ corresponds to an acoustic resonance with a wave vector $K = k_p + k_s$. This Brillouin gain can be very efficient in single-mode fibers. For example, for light at 1.5 μm the frequency shift Ω is about 11 GHz with a bandwidth of about 20 MHz. The effective bandwidth can be broadened by dithering the pump laser frequency or by using special fibers. In a recent experiment, *Tkach* and *Chraplyvy* [6.53] used Brillouin gain to selectively amplify one channel in a dense wavelength multiplexed signal by about 30 dB. The power level of the pump can be in the milliwatt range, due to the very efficient gain process in long fibers.

Another efficient gain process in fibers occurs in the anomalous dispersion regime, i.e. for wavelengths above 1.3 μm in usual silica fibers. This process, known as modulational instability, is a four-photon process that appears as an exponential growth of modulation on a cw signal. Recently, *Islam* et al. demonstrated an all-optical switch based on modulational instability [6.54]. In this device, a weak signal from a laser diode imposed modulation on a stronger pump wave. This modulation is amplified in the fiber. The experiment is performed in what amounts to a nulled interferometer, so that the modulation appears as a strong signal due to the imbalance of the interferometer.

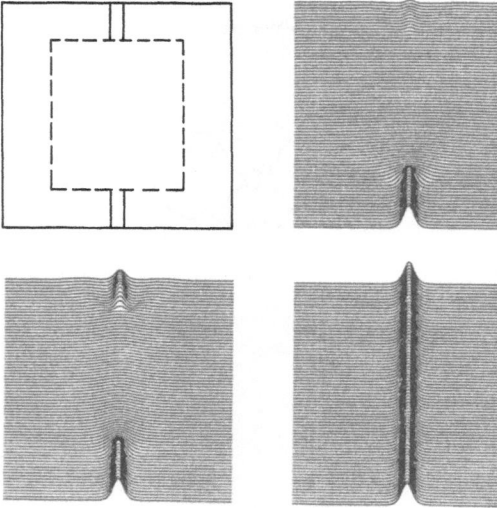

Fig. 6.14a. Self-trapping devices in nonlinear films. A nonlinear planar wave guide (region bounded by the dashed line) is embedded in a linear wave guide structure. The simulations show the beam intensity profile as it propagates in the structure from the bottom to the top. (a) A simple self-switching device based on the formation of a self-trapped beam. Simulations are shown for three power values: $0.1P_s, 0.5P_s$ and $1P_s$, where P_s is the power required to support a self-trapped beam identical to the launched mode

Fig. 6.14b. See opposite page

6.4.2 Self-Trapping Devices

In all the devices mentioned so far, the nonlinear index change is small enough that it can be considered as a perturbation over the index profile that determines the waveguide structure. These nonlinear devices operate due to small changes in the propagation constant of the mode, and the mode shape is assumed to remain unchanged.

An exciting regime lies beyond this assumption. In an extreme case, one may consider waveguides that are formed solely by the light itself. These are self-trapped beams, and they are usually discussed in the context of self-focusing [6.55]. While self-focusing effects are unstable in 3-dimensional optics, they are stable when the light is confined to a plane in a quasi 2-dimensional propagation. A self-trapped beam is a spatial soliton. The stability of solitons in the time domain is well established. Recently, a first demonstration of higher-order spatial solitons formed in nonlinear films was reported [6.56, 57]. Solitons also appear in numerical studies of high-intensity propagation in waveguides, and they can be emitted or trapped by waveguides [6.58].

To gain some insight into the possibilities of self-trapping devices, consider the simulations shown in Fig. 6.14. The structures shown assume linear channel waveguides connected by a nonlinear planar region. The propagation in the nonlinear waveguide changes from diffractive to self-trapped as a function of intensity. A simple switch based on this transition is shown in Fig. 6.14a. In Fig. 6.14b, we show how the interaction between solitons can be used to form an AND gate.

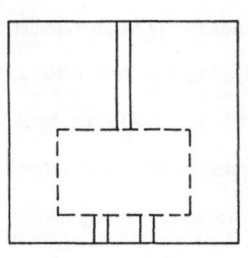
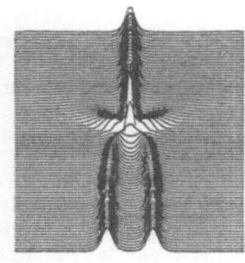

Fig. 6.14b. Interaction between two in-phase spatial solitons as an AND gate. The power of both beams is either 0 or P_s

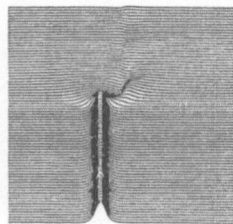
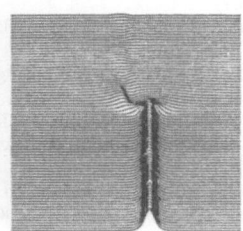

6.5 Conclusion

In this chapter, we have reviewed the state of the art of all-optical guided wave devices, and presented some of the underlying theory of operation. Clearly, the field is evolving rapidly, with many exciting ideas that have yet to be fully explored.

Before significant application can be expected, however, it will be necessary to develop nonlinear optical materials with suitable low loss and high nonlinearity characteristics. Ideally, such materials will also permit the integration of optical and electronic components. If devices based on these materials can be developed, fast all-optical switching and signal processing will play a key role in future communications and computing systems.

References

6.1 P.W. Smith: "On the Role of Photonic Switching in Future Communications Systems", in Proc. of the IEEE Int. Conf. on Communications (1987) pp. 1570–1574

6.2 A.M. Weiner, J.P. Heritage, J.A. Salehi: Encoding and decoding of femtosecond pulses. Opt. Lett. **13**, 300–302 (1988)

6.3 P.R. Prucnal, D.J. Blumenthal, P.A. Perrier: "Self-Routing Optical Switch with Optical Processing", in *Photonic Switching*, Springer Ser. Electron. Photon., Vol. 25, ed. by T.K. Gustafson, P.W. Smith (Springer, Berlin, Heidelberg 1988)

6.4 S.R. Friberg, P.W. Smith: Nonliner optical glasses for ultrafast optical switches. IEEE J. QE-**23**, 2089-2092 (1987)

6.5 G.I. Stegeman, E.M. Wright, N. Finlayson, R. Zanoni, C.T. Seaton: Third order nonlinear integrated optics, IEEE J. Lightwave Tech. **6**, 953–970 (1988)

6.6 H. Kawaguchi: Proposal for a new all-optical waveguide functional device. Opt Lett. **10**, 411–413 (1985);

L. Thylen, N. Finlayson, C.T. Seaton, G.I. Stegeman: All optical guided wave Mach-Zehnder switching devices. Appl. Phys. Lett. **51**, 1304–1306 (1987)

6.7   A. Lattes, H.A. Haus, F.J. Leonberger, E.P. Ippen: An ultrafast all-optical gate. IEEE J. QE-**19**, 1718–1723 (1983)

6.8   M. Duguay: "The Ultrafast Optical Kerr Shutter", in *Progress in Optics*, Vol. XIV, ed. by E. Wolf (North-Holland, Amsterdam 1977) pp. 161–193

6.9   R.H. Stolen, J. Botineau, A. Ashkin: Intensity discrimination of optical pulses with birefringent fiber. Opt. Lett. **7**, 512–514 (1982)

6.10   T. Morioka, M. Saruwatari, A. Takeda: Ultrafast optical multi/demultiplexer utulising optical Kerr effect in polarisation-maintaining single-mode fibres. Electron. Lett. **23**, 453–454 (1987)

6.11   N.J. Halas, D. Krokel, D. Grischkowsky: Ultrafast light-controlled optical-fiber modulator. Appl. Phys. Lett. **50**, 886–888 (1987)

6.12   Y. Silberberg, G.I. Stegeman: Nonlinear coupling of waveguide modes. Appl. Phys. Lett. **50**, 801–803 (1987)

6.13   H.G. Park, C.C. Pohalski, B.Y. Kim: Optical Kerr switch using elliptical-core two-mode fiber. Opt. Lett. **13**, 776–779 (1988)

6.14   P. Li Kam Wa, P.N. Robson, J.S. Roberts, M.A. Pate, J.P.R. David: All-optical switching between modes of a GaAs/GaAlAs multiple quantum well waveguide. Appl. Phys. Lett. **52**, 2013–2015 (1988)

6.15   M. Papuchon, A. Roy, D.B. Ostrowsky: Appl. Phys. Lett. **31**, 266 (1977)

6.16   S.M. Jensen: The nonlinear coherent coupler. IEEE J. QE-**18**, 1580–1583 (1982)

6.17   K. Kitayama, S. Wang: Optical pulse compression by nonlinear coupling. Appl. Phys. Lett. **43**, 17–19 (1983)

6.18   B. Daino, G. Gregori, S. Wabnitz: Stability analysis of nonlinear coherent coupling. J. Appl. Phys. **58**, 4512–4514 (1985)

6.19   S. Wabnitz, E.M. Wright, C.T. Seaton, G.I. Stegeman: Instabilities and all-optical phase-controlled switching in a nonlinear directional coupler. Appl. Phys. Lett. **49**, 838–840 (1986)

6.20   L. Thylen, E.M. Wright, G.I. Stegeman, C.T. Seaton, J.V. Moloney: Beam propagation analysis of a nonlinear directional coupler. Opt. Lett. **11**, 739–741 (1986)

6.21   A.A. Maier: Switching of radiation in tunnel-coupled optical waveguides by weak radiation frequency. Sov. J. Quantum Electron. **16**, 892–897 (1986)

6.22   G.I. Stegeman, C.T. Seaton, A.C. Walker, C.N. Ironside, T.J. Cullen: Nonlinear directional coupler with integrating nonlinearities. Opt. Commun. **61**, 277–281 (1987)

6.23   G.I. Stegeman, C.T. Seaton, C.N. Ironside, T.J. Cullen, A.C. Walker: Effects of saturation and loss on nonlinear directional couplers. Appl. Phys. Lett. **50**, 1035–1037 (1987)

6.24   E. Caglioti, S. Trillo, S. Wabnitz, B. Daino, G.I. Stegeman: Power dependent switching in a coherent nonlinear directional coupler in the presence of saturation. Appl. Phys. Lett. **51**, 293–295 (1987)

6.25   S. Trillo, S. Wabnitz, E. Caglioti, G.I. Stegeman: Parameter trade-offs in nonlinear directional couplers: two-level saturable non-linear media. Opt. Commun. **63**, 281–284 (1987)

6.26   S. Trillo, S. Wabnitz: Nonlinear nonreciprocity in a coherent mismatched directional coupler. Appl. Phys. Lett. **49**, 752–754 (1986)

6.27   W.M. Gibbons, D. Sarid: Model of a nonlinear directional coupler in gallium arsenide. Appl. Phys. Lett. **51**, 403–405 (1987)

6.28   E. Caglioti, S. Trillo, S. Wabnitz, G.I. Stegeman: Limitations to all-optical switching using nonlinear couplers in the presence of linear and nonlinear absorption and saturation. J. Opt. Soc. Am. B **5**, 472–482 (1988)

6.29   D.R. Heatley, E.M. Wright, J. Ehrlich, G. I. Stegeman: Nonlinear directional coupler with a diffusive Kerr-type nonlinearity. Opt. Lett. **13**, 419–421 (1988)

6.30   L. Thylen, E.M. Wright, G.I. Stegeman: A numerical analysis of nonlinear coherent couplers exhibiting saturable index changes. J. Opt. Soc. Am. B **5**, 467–471 (1988)

6.31   D.D. Gusovskii, E.M. Dianov, A.A. Maïr, V.B. Nestreuev, E.I. Shklovsii, I.A. Shcherbakov: Nonlinear light transfer in tunnel-coupled optical waveguides. Sov. J. Quantum Electron. **15**, 1523–1526 (1985)

6.32   D.D. Gusovskii, E.M. Dianov, A.A. Maïr, V.B. Nestreuev, V.V. Osiko, A.M. Prokhorov, K. Yu. Sitarskii, I.A. Shcherbakov: Experimental observation of the self-switching of radiation in tunnel-coupled optical waveguides. Sov. J. Quantum Electron. **17**, 724–727 (1987)

6.33   A.A. Maïr, Yu. N. Serdyuchenko, K. Yu. Sitarskii, M. Ya. Shchelev, A.I. Shcherbakov: Breakup of an ultrashort pulse in the course of self-switching of light in tunnel-coupled waveguides. Sov. J. Quantum Electron. **17**, 735–736 (1987)

6.34   S.R. Friberg, Y. Silberberg, M.K. Oliver, M.J. Andrejco, M.A. Saifi, P.W. Smith: Ultrafast all-optical switching in dual-core fiber nonlinear coupler. Appl. Phys. Lett. **51**, 1135–1137 (1987)

6.35   S.R. Friberg, A.M. Weiner, Y. Silberberg, G. Sfez, P.W. Smith: Femtosecond switching in a dual-core-fiber nonlinear coupler. Opt. Lett. **13**, 904–906 (1988)

6.36   A.A. Maier: Possible practical use of self-switching of radiation in coupled waveguides for amplification of the useful modulation of a signal. Sov. J. Quantum Electron. **17**, 1013–1017 (1987)

6.37   P. Li Kam Wa, J.E. Stich, N.J. Mason, J.S. Roberts, P.N. Robson: All-optical multiple-quantum-well waveguide switch. Electron. Lett. **21**, 26–27 (1985)

6.38   M. Cada, R.C. Gauthier, B.E. Paton, J. Chrostowski: Nonlinear guided waves coupled nonlinearity in a planar GaAs/GaAlAs multiple quantum well structure. Appl. Phys. Lett. **49**, 755–757 (1986)
        M. Cada, B.P. Keyworth, J.M. Glinski, A.J. Spring Thorpe, P. Mandeville: Experiment with multiple-quantum-well waveguide switching element. J. Opt. Soc. Am. B **5**, 462–466 (1988)

6.39   B. Daino, G. Gregori, S. Wabnitz: New all-optical devices based on third-order nonlinearity of birefringent fibers. Opt. Lett. **11**, 42–44 (1986)

6.40   R. Jin, C.L. Chuang, H.M. Gibbs, S.W. Koch, J.N. Polky, G.A. Pubanz: Picosecond all-optical switching in single mode GaAs/AlGaAs strip-loaded nonlinear directional couplers. Appl. Phys. Lett. **53**, 1791–1793 (1988)

6.41   S. Trillo, S. Wabnitz, R.H. Stolen, G. Assanto, C.T. Seaton, G.I. Stegeman: Experimental observation of polarization instability in a birefringent optical fiber. Appl. Phys. Lett. **49**, 1224–1226 (1986)

6.42   S. Trillo, S. Wabnitz, N. Finlayson, W.C. Banyai, C.T. Seaton, G.I. Stegeman, R.H. Stolen: Picosecond nonlinear polarization switching with a fiber filter. Appl. Phys. Lett. **53**, 837–839 (1988)

6.43   W.K. Burns, A.F. Milton: Mode conversion in planar dielectric separating waveguides. IEEE QE-11, 32–39 (1975); An analytic solution for mode coupling in optical waveguide branches. IEEE J. QE-16, 446–454 (1980)

6.44   Y. Silberberg, P. Perlmutter, J.E. Baran: Digital optical switch. Appl. Phys. Lett. **51**, 1230–1232 (1987)

6.45   Y. Silberberg, B.G. Sfez: All-optical phase- and power-controlled switching in nonlinear waveguide junctions. Opt. Lett. **13**, 1132–1134 (1988)

6.46   A.M. Weiner, J.P. Heritage, E.M. Kirschner: High-resolution femtosecond pulse shaping. J. Opt. Soc. Am. B **5**, 1563–1570 (1988)

6.47   Y. Silberberg, A.M. Weiner, H. Fouckhardt, D.E. Laeird, M.A. Saifi, M.J. Andrejco, P.W. Smith: "Improved Operation of a Two-Core Fiber All-Optical Switch using Femtosecond Square Optical Pulses", Topical Meeting on Photonic Switching, Salt Lake City, Utah, March 1989

6.48   M. Shirasaki, H.A. Haus, D. Liu Wong: In Digest of Conference on Lasers and Electro-Optics (OSA, Washington, DC 1987) paper THO1

6.49   N.J. Doran, D. Wood: A soliton processing element for all-optical switching and logic. J. Opt. Soc. Am. B **4**, 1843–1846 (1987); Nonlinear optical loop mirror. Opt. Lett. **13**, 56–58 (1988)

6.50   K.J. Blow, N.J. Doran, B.K. Nayar: Experimental demonstration of optical soliton switching in an all-fibre nonlinear Sagnac interferometer. Opt. Lett. **14**, 754–756 (1989)

6.51   M.N. Islam, E.R. Sunderman, R.H. Stolen, W. Pleibel, J.R. Simpson: "Soliton Switching in a Fiber Nonlinear Loop Mirror", in Topical Meeting on Photonics Switching, Salt Lake City, Utah, March 1989, paper PD-14

6.52   S. Trillo, S. Wabnitz, E.M. Wright, G.I. Stegeman: Soliton switching in fiber nonlinear directional couplers. Opt. Lett. **13**, 672–674 (1988)

6.53   R.W. Tkach, A.R. Chraplyvy: "Nonlinear Optical Star Network Using Tunable Narrowband Brillouin Filters", International Conference on Quantum Electronics, IQEC 1988, Tokyo, Japan, Paper WG1

6.54   M.N. Islam, S.P. Dijaili, J.P. Gordon: Modulation-instability-based fiber interferometer switch near 1.5 μm. Opt. Lett. **13**, 518–520 (1988)

6.55   R. Y. Chiao, E. Garmire, C.H. Townes: Self-trapping of optical beams. Phys. Rev. Lett. **13**, 479–482 (1964)

6.56   S. Maneuf, F. Reynaud: Quasi-steady self-trapping of first, second and third order sub-nanosecond soliton beams. Opt. Commun. **66**, 325–328 (1988)

6.57   S. Maneuf, R. Desailly, C. Froehly: Stable self-trapping of laser beams: observation in a nonlinear planar waveguide. Opt. Commun. 65, 193–198 (1988)

6.58   D.R. Heatley, E.M. Wright, G.I. Stegeman: Soliton coupler. Appl. Phys. Lett. 53, 172–174 (1988)

Subject Index